2017

Guide to Occupational Exposure Values

Compiled by
ACGIH®

ACGIH®

Defining the Science of
Occupational and Environmental Health®

Signature Publications

The *Guide to Occupational Exposure Values* is a readily accessible reference for comparison of published values from ACGIH®; the U.S. Occupational Safety and Health Administration (OSHA); the U.S. National Institute for Occupational Safety and Health (NIOSH); Deutsche Forschungsgemeinschaft (DFG), Federal Republic of Germany, Commission for the Investigation of Health Hazards of Chemical Compounds in the Work Area; and the American Industrial Hygiene Association (AIHA). Provided below are the sources of the values cited in this *Guide*, including publication dates, and the uniform resource locator (URL) if verified online. November 2016 was the date of online verification.

- ACGIH® Threshold Limit Values (TLVs®) for Chemical Substances

  - *2017 TLVs® and BEIs®: Threshold Limit Values for Chemical Substances and Physical Agents and Biological Exposure Indices.* ACGIH®, Cincinnati, OH (2017).

- OSHA Permissible Exposure Limits (PELs)

  - Title 29, Code of Federal Regulations, Part 1910.1000-1910.1200, Air Contaminants, Final Rule, specified in Tables Z-1, Z-2, and Z-3; Federal Register 58:35338-35351, June 30, 1993; corrected in Federal Register 58:40191, July 27, 1993; amended in Federal Register 60:9624, February 21, 1995; Federal Register 60:33343, June 28, 1995; corrected in Federal Register 60:33984, June 29, 1995; Federal Register 62:42018, August 4, 1997; and subsequent corrections/amendments/proposals through Federal Register 71:10373, February 28, 2006. Reviewed at http://www.osha.gov/pls/oshaweb/owadisp.show_document?p_table=STANDARDS&p_id=9992.

- NIOSH Recommended Exposure Limits (RELs)

  - NIOSH Pocket Guide to Chemical Hazards: Introduction. Available online at: http://www.cdc.gov/niosh/npg/pgintrod.html (Reviewed 2016).

  - *See also*: Ludwig HR; Cairelli SG; Whalen JJ (Eds): Documentation for Immediately Dangerous to Life or Health Concentrations (IDLH): Introduction. NTIS Pub. No. PB-94-195047 (1994). Available online at: http://www.cdc.gov/niosh/idlh/idlhintr.html.

  - *See also*: Wittaker C; Rice F; McKernan L; et al.: Current intelligence bulletin 68: NIOSH chemical carcinogen policy. Department of Health and Human Services, Centers for Disease Control and Prevention, National Institute for Occupational Safety and Health (NIOSH), DHHS, Publication No. 2017-100 (2017). Available online at: https://www.cdc.gov/niosh/docs/2017-100/pdf/2017-100.pdf.

- DFG Maximum Concentrations at the Workplace (MAKs)

  - List of MAK and BAT Values 2016: Maximum Concentrations and Biological Tolerance Values at the Workplace. Report No. 52. Commission for the Investigation of Health Hazards of Chemical Compounds in the Work Area. Wiley-VCH Verlag GmbH & Co. KGaA, Weinheim, FRG (September 12, 2016).

- AIHA Workplace Environmental Exposure Levels (WEELs™)

  - 2011 current WEEL® values. AIHA Guideline Foundation, Fairfax, VA. Available online at http://www.aiha.org/get-involved/AIHAGuidelineFoundation/WEELS/Documents/2011WEELValues.pdf.

- OARS Workplace Environmental Exposure Levels (WEELs™)

  ○ Occupational Alliance for Risk Science (OARS) Workplace Environmental Exposure Levels (WEELs™) managed by Toxicological Excellence for Risk Assessment (TERA), Cincinnati, OH. Available online at http://www.tera.org/OARS/WEEL.html (Reviewed 2016).

The *Guide* also includes those carcinogens found in the occupational environment that are identified by the above organizations and by the U.S. Environmental Protection Agency (EPA), the International Agency for Research on Cancer (IARC), and the U.S. National Toxicology Program (NTP). In addition to those sources cited above, the following were also used in preparing this *Guide* and were reviewed in November 2016.

- U.S. EPA Integrated Risk Information System (IRIS) database. A–Z List of Substances. Online at: http://cfpub.epa.gov/ncea/iris/index.cfm?fuseaction=iris.showsubstancelist&list_type =alpha&view=all.

- Agents Classified by the IARC Monographs, Volumes 1–117. IARC, Lyon, France (1987–2016). Available online at: http://monographs.iarc.fr/ENG/classification/latest_classif.php (Reviewed 2016).

- Report on Carcinogens, 14th Ed., U.S. Department of Health and Human Services, Public Health Service, National Toxicology Program, Research Triangle Park, NC (2016). Available online at: http://ntp.niehs.nih.gov/pubhealth/roc/index-1.html (Reviewed December 2016).

The *Guide to Occupational Exposure Values* is intended as a companion document to the ACGIH® annual *Threshold Limit Values for Chemical Substances and Physical Agents and Biological Exposure Indices* (*TLVs® and BEIs®*) book, specifically the section on TLVs® for Chemical Substances in the Work Environment.

The following pages provide "Definitions, Abbreviations, Terms, and Coding," the MAK "Peak Exposure Limitation Categories," the MAK "Pregnancy Risk Group Classifications," and the MAK "Germ Cell Mutagens Classifications."

Editor's note: The double entries that were previously included in this publication were eliminated effective with the 2006 edition. The entry in this publication will correspond to that carried in the *TLVs® and BEIs®* book, e.g., 2-butoxyethanol rather than ethylene glycol monobutyl ether. When ACGIH® does not recommend a TLV® and two or more jurisdictions (e.g., MAK and IARC) list a chemical substance with separate synonyms, ACGIH® will generally use the ChemIDplus database available on the ToxNet website (http://toxnet.nlm.nih.gov/) maintained by the U.S. National Library of Medicine. ChemIDplus is a database of over 370,000 chemicals, which contains names and synonyms as well as chemical formulae and structures. Whichever synonym ChemIDplus uses as the primary name attached to a specific CAS number is the name generally listed in this publication. In all cases, the removed synonym is listed with its primary entry and with its respective CAS number in the CAS Number Index section of this publication.

Carcinogenicity Categories

U.S. Environmental Protection Agency (EPA)

NOTE: The rationale and methods used to develop the carcinogenicity classifications EPA-A through EPA-E are found in the 1986 *Risk Assessment Guidelines* (EPA/600/8-87/045). The categories, EPA-K, EPA-L, EPA-CBD, and EPA-UL, were developed under the 1996 *Proposed Guidelines for Carcinogen Risk Assessment* (*Federal Register* 61[79]:17960-18011, April 23, 1996). Further to its updating of risk assessment guidelines, EPA issued a revised draft *Guidelines for Carcinogen Risk Assessment* (NCEA-F-0644; July 1999), which resulted in slightly different descriptors. In 2005, the agency published the final version of *Guidelines for Carcinogen Risk Assessment* (EPA/630/P-03-001 B), which contained refined descriptors for summarizing weight of evidence for human carcinogenic potential. All four risk assessment guidelines may be found online at: **http://www.epa.gov/risk**. In all instances, the user is referred to the online IRIS Guidance Documents found on the EPA website: **http://www.epa.gov/iris** and the online Toxicological Reviews and Support Documents available at: **http://cfpub.epa.gov/ncea/iris** for further carcinogenicity discussion and for information on long-term toxic effects other than carcinogenicity. In all cases, the most current carcinogenicity assessment will be listed in this publication.

EPA-A: Human Carcinogen — Sufficient evidence from epidemiologic studies to support a causal association between exposure and cancer.

-B: Probable Human Carcinogen — Weight of evidence of human carcinogenicity based on epidemiologic studies is limited; agents for which weight of evidence of carcinogenicity based on animal studies is sufficient.

Two subgroups:

-B1: Limited evidence of carcinogenicity from epidemiologic studies.

-B2: Sufficient evidence from animal studies; inadequate evidence or no data from epidemiologic studies.

-C: Possible Human Carcinogen — Limited evidence of carcinogenicity in animals in the absence of human data.

-D: Not Classifiable as to Human Carcinogenicity — Inadequate human and animal evidence of carcinogenicity or no data are available.

-E: Evidence of Noncarcinogenicity for Humans — No evidence for carcinogenicity in at least two adequate animal tests in different species or in both adequate epidemiologic and animal studies.

Under the 1996 Draft Guidelines, when the available tumor effects and other key data are adequate to demonstrate carcinogenic potential convincingly for humans, EPA-K or EPA-L are appropriate descriptors.

EPA-K: Known Human Carcinogens — Agents *known* to be carcinogenic in humans based on either epidemiologic evidence or a combination of epidemiologic and experimental evidence, demonstrating causality between human exposure and cancer;

OR

Agents that should be treated *as if* they were *known* human carcinogens, based on a combination of epidemiologic data showing a plausible causal association (not demonstrating it definitively) and strong experimental evidence.

-L: Likely to Produce Cancer in Humans — Agents that are *likely* to produce cancer in humans due to the production or anticipated production of tumors by modes of action that are relevant or assumed to be relevant to human carcinogenicity. Modifying descriptors for particularly high or low ranking in the "known/likely" group can be applied based on scientific judgment and experience and are as follows:

- Agents that are *likely* to produce cancer in humans based on data that are at the high end of the weights of evidence typical of this group.

- Agents that are *likely* to produce cancer in humans based on data that are at the low end of the weights of evidence typical of this group.

-CBD: Cannot Be Determined — This descriptor is appropriate when available tumor effects or other key data are suggestive or conflicting or limited in quantity and, thus, are not adequate to convincingly demonstrate carcinogenic potential for humans. In general, further agent specific and generic research and testing are needed to be able to describe human carcinogenic potential. The descriptor *cannot be determined* is used with a subdescriptor that captures the rationale:

- Agents whose carcinogenic potential *cannot be determined*, but for which there is suggestive evidence that raises concern for carcinogenic effects.

- Agents whose carcinogenic potential cannot be determined because the existing evidence is composed of *conflicting data* (e.g., some evidence is suggestive of carcinogenic effects, but other equally pertinent evidence does not confirm any concern).

- Agents whose carcinogenic potential *cannot be determined* because there are inadequate data to perform an assessment.

- Agents whose carcinogenic potential *cannot be determined* because no data are available to perform an assessment.

-NL: Not Likely to be Carcinogenic in Humans — This descriptor is appropriate when experimental evidence is satisfactory for deciding that there is no basis for human hazard concern, as follows (in the absence of human data suggesting a potential for cancer effects):

- Agents *not likely* to be carcinogenic to humans because they have been evaluated in at least two well-conducted studies in two appropriate animal species without demonstrating carcinogenic effects.

- Agents *not likely* to be carcinogenic to humans because they have been appropriately evaluated in animals and show only carcinogenic effects that have been shown not to be relevant to humans (e.g., showing only effects in the male rat kidney due to accumulation of α_2u-globulin).

- Agents *not likely* to be carcinogenic to humans when carcinogenicity is dose or route dependent. For instance, not likely below a certain dose range (categorized as *likely* above that range) or *not likely* by a certain route of exposure (may be categorized as likely by another route of exposure). To qualify, agents will have been appropriately evaluated in animal studies and the only effects show a dose range or route limitation or a route limitation is otherwise shown by empirical data.

Under the 1999 revised draft Guidelines, the following descriptors were issued; however, the descriptors are only presented in the context of a weight-of-evidence-narrative. [*Editor's note*: The "short hand" used within this *Guide* (e.g., EPA-K) to indicate descriptors used within the 1996 and 1999 draft Guidelines were developed to accommodate the page format only.] The reader is referred to the current EPA evaluation for a complete discussion of substance in question.

EPA-CaH: Carcinogenic to Humans — This descriptor is appropriate when there is convincing epidemiologic evidence demonstrating causality between human exposure and cancer. This descriptor is also appropriate when there is an absence of conclusive epidemiologic evidence to clearly establish a cause and effect relationship between human exposure and cancer, but there is compelling evidence of carcinogenicity in animals and mechanistic information in animals and humans demonstrating similar mode(s) of carcinogenic action. It is used when all of the following conditions are met:

- There is evidence in a human population(s) of association of exposure to the agent with cancer, but not enough to show a causal association;

- There is extensive evidence of carcinogenicity;

- The mode(s) of carcinogenic action and associated key events have been identified in animals; and

- The key events that precede the cancer response in animals have been observed in the human population(s) that also show evidence of an association of exposure to the agent with cancer.

-L: Likely to be Carcinogenic to Humans — This descriptor is appropriate when the available tumor effects and other key data are adequate to demonstrate carcinogenic potential to humans. Adequate data are within a spectrum. At one end is evidence for an association between human exposure to the agent and cancer and strong experimental evidence of carcinogenicity in animals; at the other, with no human data, the weight of experimental evidence shows animal carcinogenicity by a mode or modes of action that are relevant or assumed to be relevant to humans.

-S: Suggestive Evidence of Carcinogenicity, but Not Sufficient to Assess Human Carcinogenic Potential — This descriptor is appropriate when the evidence from human or animal data is suggestive of carcinogenicity, which raises a concern for carcinogenic effects but is judged not sufficient for a conclusion as to human carcinogenic potential. Examples of such evidence may include: a marginal increase in tumors that may be exposure-related, or evidence is observed only in a single study, or the only evidence is limited to certain high background tumors in one sex of one species. Dose–response assessment is not indicated for these agents. Further studies would be needed to determine human carcinogenic potential.

-I: Data are Inadequate for an Assessment of Human Carcinogenic Potential — This descriptor is used when available data are judged inadequate to perform an assessment. This includes a case when there is a lack of pertinent or useful data or when existing evidence is conflicting, e.g., some evidence is suggestive of carcinogenic effects, but other equally pertinent evidence does not confirm a concern.

-NL: Not Likely to be Carcinogenic to Humans — This descriptor is used when the available data are considered robust for deciding that there is no basis for human hazard concern. The judgment may be based on the following:

- Extensive human experience that demonstrates lack of carcinogenic effect (e.g., phenobarbital).

- Animal evidence that demonstrates lack of carcinogenic effect in at least two well-designed and well-conducted studies in two appropriate animal species (in the absence of human data suggesting a potential for cancer effects).

- Extensive experimental evidence showing that the only carcinogenic effects observed in animals are not considered relevant to humans (e.g., showing only effects in the male rat kidney due to accumulation of α_2u-globulin).

- Evidence that carcinogenic effects are not likely by a particular route of exposure (Section 2.3.3.).

- Evidence that carcinogenic effects are not anticipated below a defined dose range.

Under the 2005 *Guidelines for Carcinogen Risk Assessment* (EPA/630/P-03/001 B), the following descriptors were issued; however, the descriptors are only presented in the context of a weight-of-evidence-narrative. [*Editor's note*: The "short hand" (e.g., EPA-CaH) used within this *Guide* to indicate the 2005 descriptors was developed to accommodate the page format of this *Guide* only.] The reader is referred to the individual IRIS evaluation for a complete discussion of the substance in question.

EPA-CaH: Carcinogenic to Humans — This descriptor indicates strong evidence of human carcinogenicity. It covers different combinations of evidence.

- This descriptor is appropriate when there is convincing epidemiologic evidence of a causal association between human exposure and cancer.

- Exceptionally, this descriptor may be equally appropriate with a lesser weight of epidemiologic evidence that is strengthened by other lines of evidence. It can be used when all of the following conditions are met: (a) there is strong evidence of an association between human exposure and either cancer or the key precursor events of the agent's mode of action but not enough for a causal association, and (b) there is extensive evidence of carcinogenicity in animals, and (c) the mode(s)of carcinogenic action and associated key precursor events have been identified in animals, and (d) there is strong evidence that the key precursor events that precede the cancer response in animals are anticipated to occur in humans and progress to tumors, based on available biological information. In this case, the narrative includes a summary of both the experimental and epidemio-logic information on mode of action and also an indication of the relative weight that each source of information carries, e.g., based on human information, based on limited human and extensive animal experiments.

-L: Likely to Be Carcinogenic to Humans — This descriptor is appropriate when the weight of the evidence is adequate to demonstrate carcinogenic potential to humans but does not reach the weight of evidence for the "Carcinogenic to Humans" descriptor. Adequate evidence consistent with this descriptor covers a broad spectrum. As stated previously, the use of the term "likely" as a weight of evidence descriptor does not correspond to a quantifiable probability. The examples below are meant to represent the broad range of data combinations that are covered by this descriptor; they are illustrative and provide neither a checklist nor a limitation for the data that might support use of this descriptor. Moreover, additional information, e.g., on mode of action, might change the choice of descriptor for the illustrated examples. Supporting data for this descriptor may include:

- an agent demonstrating a plausible (but not definitively causal) association between human exposure and cancer, in most cases with some supporting biological, experimental evidence, though not necessarily carcinogenicity data from animal experiments;

- an agent that has tested positive in animal experiments in more than one species, sex, strain, site, or exposure route, with or without evidence of carcinogenicity in humans;

- a positive tumor study that raises additional biological concerns beyond that of a statistically significant result, for example, a high degree of malignancy, or an early age at onset;

- a rare animal tumor response in a single experiment that is assumed to be relevant to humans; or

- a positive tumor study that is strengthened by other lines of evidence, for example, either plausible (but not definitively causal) association between human exposure and cancer <u>or</u> evidence that the agent or an important metabolite causes events generally known to be associated with tumor formation (such as DNA reactivity or effects on cell growth control) likely to be related to the tumor response in this case.

-S: Suggestive Evidence of Carcinogenic Potential — This descriptor of the database is appropriate when the weight of evidence is suggestive of carcinogenicity; a concern for potential carcinogenic effects in humans is raised, but the data are judged not sufficient for a stronger conclusion. This descriptor covers a spectrum of evidence associated with varying levels of concern for carcinogenicity, ranging from a positive cancer result in the only study on an agent to a single positive cancer result in an extensive database that includes negative studies in other species. Depending on the extent of the database, additional studies may or may not provide further insights. Some examples include:

- a small, and possibly not statistically significant, increase in tumor incidence observed in a single animal or human study that does not reach the weight of evidence for the descriptor "Likely to Be Carcinogenic to Humans." The study generally would not be contradicted by other studies of equal quality in the same population group or experimental system (*see* discussions of *conflicting evidence* and differing results, below);

- a small increase in a tumor with a high background rate in that sex and strain, when there is some but insufficient evidence that the observed tumors may be due to intrinsic factors that cause background tumors and not due to the agent being assessed. (When there is a high background rate of a specific tumor in animals of a particular sex and strain, then there may be biological factors operating independently of the agent being assessed that could be responsible for the development of the observed tumors.) In this case, the reasons for determining that the tumors are not due to the agent are explained;

- evidence of a positive response in a study whose power, design, or conduct limits the ability to draw a confident conclusion (but does not make the study fatally flawed), but where the carcinogenic potential is strengthened by other lines of evidence (such as structure-activity relationships); or

- a statistically significant increase at one dose only, but no significant response at the other doses and no overall trend.

-II: Inadequate Information to Assess Carcinogenic Potential — This descriptor of the database is appropriate when available data are judged inadequate for applying one of the other descriptors. Additional studies generally would be expected to provide further insights. Some examples include:

- little or no pertinent information;

- conflicting evidence, that is, some studies provide evidence of carcinogenicity but other studies of equal quality in the same sex and strain are negative. *Differing results*, that is, positive results in some studies and negative results in one or more different experimental systems, do not constitute *conflicting evidence*, as the term is used here. Depending on the overall weight of evidence, differing results can be considered either suggestive evidence or likely evidence; or

- negative results that are not sufficiently robust for the descriptor, "Not Likely to Be Carcinogenic to Humans."

-NL: Not Likely to Be Carcinogenic to Humans — This descriptor is appropriate when the available data are considered robust for deciding that there is no basis for human hazard concern. In some instances, there can be positive results in experimental animals when there is strong, consistent evidence that each mode of action in experimental animals does not operate in humans. In other cases, there can be convincing evidence in both humans and animals that the agent is not carcinogenic. The judgment may be based on data such as:

- animal evidence that demonstrates lack of carcinogenic effect in both sexes in well-designed and well-conducted studies in at least two appropriate animal species (in the absence of other animal or human data suggesting a potential for cancer effects);

- convincing and extensive experimental evidence showing that the only carcinogenic effects observed in animals are not relevant to humans;

- convincing evidence that carcinogenic effects are not likely by a particular exposure route; or

- convincing evidence that carcinogenic effects are not likely below a defined dose range.

The "Not Likely" descriptor applies only to the circumstances supported by the data. For example, an agent may be "Not Likely to Be Carcinogenic" by one route but not necessarily by another. In those cases that have positive animal experiment(s) but the results are judged to be not relevant to humans, the narrative discusses why the results are not relevant.

International Agency for Research on Cancer (IARC)

IARC-1: Carcinogenic to Humans — The exposure circumstance entails exposures that are carcinogenic to humans. This category is used when there is *sufficient evidence* of carcinogenicity in humans. Exceptionally, an agent (mixture) may be placed in the category when evidence in humans is less than sufficient but there is *sufficient evidence* of carcinogenicity in experimental animals and strong evidence in exposed humans that the agent (mixture) acts through a relevant mechanism of carcinogenicity.

-2A: Probably Carcinogenic to Humans — The exposure circumstance entails exposures that are probably carcinogenic to humans. This category is used when there is *limited evidence* of carcinogenicity in humans and sufficient evidence of carcinogenicity in experimental animals. In some cases, an agent (mixture) may be classified in this category when there is inadequate evidence of carcinogenicity in humans and *sufficient evidence* of carcinogenicity in experimental animals and strong evidence that the carcinogenesis is mediated by a mechanism that also operates in humans. Exceptionally, an agent, mixture, or exposure circumstance may be classified in this category solely on the basis of limited evidence of carcinogenicity in humans.

-2B: Possibly Carcinogenic to Humans — The exposure circumstance entails exposures that are possibly carcinogenic to humans. This category is used for agents, mixtures, and exposure circumstances for which there is *limited evidence* of carcinogenicity in humans and less than *sufficient evidence* of carcinogenicity in experimental animals. It may also be used when there is *inadequate evidence* of carcinogenicity in humans but there is *sufficient evidence* of carcinogenicity in experimental animals. In some instances, an agent, mixture, or exposure circumstance for which there is inadequate evidence of carcinogenicity in humans but *limited evidence* of carcinogenicity in experimental animals together with supporting evidence from other relevant data may be placed in the group.

-3: Unclassifiable as to Carcinogenicity in Humans — This category is used most commonly for agents, mixtures, and exposure circumstances for which the evidence of carcinogenicity is inadequate in humans and inadequate or limited in experimental animals. Exceptionally, agents (mixtures) for which the evidence of carcinogenicity is inadequate in humans but sufficient in experimental animals may be placed in this category when there is strong evidence that the mechanism of carcinogenicity in experimental animals does not operate in humans. Agents, mixtures, and exposure circumstances that do not fall into any other group are also placed in this category.

-4: Probably Not Carcinogenic to Humans — This category is used for agents or mixtures for which there is *evidence suggesting lack of carcinogenicity* in humans and in experimental animals. In some instances, agents or mixtures for which there is *inadequate evidence* of carcinogenicity in humans but *evidence suggesting lack of carcinogenicity* in experimental animals, consistently and strongly supported by a broad range of other relevant data, may be classified in this group.

German MAK Commission

MAK-1: Substances that cause cancer in man and can be assumed to make a significant contribution to cancer risk. Epidemiological studies provide adequate evidence of a positive correlation between the exposure of humans and the occurrence of cancer. Limited epidemiological data can be substantiated by evidence that the substance causes cancer by a mode of action that is relevant to man.

-2: Substances that are considered to be carcinogenic for man because sufficient data from long-term animal studies or limited evidence from animal studies substantiated by evidence from epidemiological studies indicate that they can make a significant contribution to cancer risk. Limited data from animal studies can be supported by evidence that the substance causes cancer by a mode of action that is relevant to man and by results of *in vitro* tests and short-term animal studies.

-3: Substances which cause concern that they could be carcinogenic for man but cannot be assessed conclusively because of lack of data. The classification in Category 3 is provisional.

-3A: Substances for which the criteria for classification in Category 4 or 5 are fulfilled but for which the database is insufficient for the establishment of a MAK value.

-3B: Substances for which *in vitro* tests or animal studies have yielded evidence of carcinogenic effects that is not sufficient for classification of the substance in one of the other categories. Further studies are required before a final classification can be made. A MAK or BAT value can be established, provided no genotoxic effects have been detected.

-4: Substances with carcinogenic potential for which genotoxicity plays no or at most a minor role. No significant contribution to human cancer risk is expected, provided the MAK value is observed. The classification is supported especially by evidence that increases in cellular proliferation or changes in cellular differentiation are important in the mode of action. To characterize the cancer risk, the manifold mechanisms contributing to carcinogenesis and their characteristic dose–time–response relationships are taken into consideration.

-5: Substances with carcinogenic and genotoxic effects, the potency of which is considered to be so low that, provided the MAK and BAT values are observed, no significant contribution to human cancer risk is to be expected. The classification is supported by information on the mode of action, dose-dependence, and toxicokinetic data pertinent to species comparison.

U.S. National Institute for Occupational Safety and Health (NIOSH)

NIOSH-Ca: Potential occupational carcinogen, with no further categorization.

U.S. National Toxicology Program (NTP)

NTP-K: Known to Be a Human Carcinogen — There is sufficient evidence of carcinogenicity from studies in humans which indicates a causal relationship between exposure to the agent, substance or mixture and human cancer.

-R: Reasonably Anticipated to Be a Human Carcinogen (RAHC) — There is limited evidence of carcinogenicity from studies in humans, which indicates that causal interpretation is credible, but that alternative explanations, such as chance, bias or confounding factors, could not adequately be excluded;

OR

There is sufficient evidence of carcinogenicity from studies in experimental animals which indicates there is an increased incidence of malignant and/or a combination of malignant and benign tumors: (1) in multiple species or at multiple tissue sites, or (2) by multiple routes of exposure, or (3) to an unusual degree with regard to incidence, site or type of tumor, or age at onset;

OR

There is less than sufficient evidence of carcinogenicity in humans or laboratory animals, however; the agent, substance or mixture belongs to a well defined, structurally-related class of substances whose members are listed in a previous Report on Carcinogens as either a known to be human carcinogen or reasonably anticipated to be human carcinogen, or there is convincing relevant information that the agent acts through mechanisms indicating it would likely cause cancer in humans.

U.S. Occupational Safety and Health Administration (OSHA)

OSHA-Ca: Carcinogen defined with no further categorization.

American Conference of Governmental Industrial Hygienists (ACGIH®)

TLV-A1: Confirmed Human Carcinogen — The agent is carcinogenic to humans based on the weight of evidence from epidemiologic studies.

-A2: Suspected Human Carcinogen — Human data are accepted as adequate in quality but are conflicting or insufficient to classify the agent as a confirmed human carcinogen; OR, the agent is carcinogenic in experimental animals at dose(s), by route(s) of exposure, at site(s), of histologic type(s), or by mechanism(s) considered relevant to worker exposure. The A2 is used primarily when there is limited evidence of carcinogenicity in humans and sufficient evidence of carcinogenicity in experimental animals with relevance to humans.

-A3: Confirmed Animal Carcinogen with Unknown Relevance to Humans — The agent is carcinogenic in experimental animals at a relatively high dose, by route(s) of administration, at site(s), of histologic type(s), or by mechanism(s) that may not be relevant to worker exposure. Available epidemiologic studies do not confirm an increased risk of cancer in exposed humans. Available evidence does not suggest that the agent is likely to cause cancer in humans except under uncommon or unlikely routes or levels of exposure.

-A4: Not Classifiable as a Human Carcinogen — Agents which cause concern that they could be carcinogenic for humans but which cannot be assessed conclusively because of a lack of data. In vitro or animal studies do not provide indications of carcinogenicity which are sufficient to classify the agent into one of the other categories.

-A5: Not Suspected as a Human Carcinogen — The agent is not suspected to be a human carcinogen on the basis of properly conducted epidemiologic studies in humans. These studies have sufficiently long follow-up, reliable exposure histories, sufficiently high dose, and adequate statistical power to conclude that exposure to the agent does not convey a significant risk of cancer to humans; OR, the evidence suggesting a lack of carcinogenicity in experimental animals is supported by mechanistic data.

Substances for which no human or experimental animal carcinogenic data have been reported are assigned no carcinogen designation.

Exposures to carcinogens must be kept to a minimum. Workers exposed to A1 carcinogens without a TLV® should be properly equipped to eliminate to the fullest extent possible all exposure to the carcinogen. For A1 carcinogens with a TLV® and for A2 and A3 carcinogens, worker exposure by all routes should be carefully controlled to levels as low as possible below the TLV®.

Notations

A–D listed in DFG MAK column only refers to Pregnancy Risk Group Classifications; *see* page xv for definitions.

(D) "Inert" gas or vapor that acts primarily as a simple asphyxiant without other significant physiologic effects when present in high concentrations in air.

DSEN May cause dermal sensitization. This notation is used to indicate the potential for dermal sensitization resulting from the interaction of an absorbed agent and ultraviolet light (i.e., photosensitization).

DSEN TLV Potential for worker sensitization by dermal contact as confirmed by available human or animal data.

E The value is for particulate matter containing no asbestos and < 1% Crystalline silica.

(EX) Explosion hazard: the substance is a flammable asphyxiant or excursions above the TLV® could approach 10% of the lower explosive limit.

(F) Respirable fibers: length > 5 μ; aspect ratio $\geq$ 3:1, as determined by the membrane filter method at 400–450x magnification (4-mm objective), using phase-contrast illumination.

G As measured by the vertical elutriator, cotton-dust sampler. *See* Cotton Dust TLV® *Documentation*.

(H) Aerosol only.

I Measured as Inhalable fraction of the aerosol.

IFV Measured as Inhalable fraction and vapor.

(J) Does not include stearates of toxic metals.

(K) Should not exceed 2 mg/m^3 respirable particulate.

L Exposure to carcinogens must be kept to a minimum. Workers exposed to A1 carcinogens without a TLV® should be properly equipped to eliminate to the fullest extent possible all exposure to the carcinogen. For A2 and A3 carcinogens without a TLV®, worker exposure by all routes should be carefully controlled. *See* the ACGIH® carcinogen definitions starting on the previous page.

(O) Sampled by method that does not collect vapor.

P Avoid prolonged and repeated skin contact to diesel fuels which can lead to dermal irritation and may be associated with an increased risk of skin cancer.

Q Absorbed rapidly through the skin in molten or heated liquid form in amounts that have caused rapid death in humans.

R Measured as respirable fraction of the aerosol.

RSEN May cause respiratory sensitization.

RSEN TLV Potential for worker sensitization by inhalation exposure as confirmed by available human or animal data.

Sa MAK–danger of sensitization of the airways.

SEN TLV Confirmed potential for worker sensitization as a result of dermal contact and/or inhalation exposure, based on the weight of scientific evidence.

Sh MAK–danger of sensitization of the skin.

Sah MAK–danger of sensitization of the airways and the skin.

SP MAK–danger of photo-contact sensitization.

T Measured as thoracic fraction of the aerosol.

(V) Vapor fraction.

Notations (continued)

(W) Worker exposure by all routes should be minimized to the fullest extent possible.

1–5 listed in the DFG MAK column only refers to Germ Cell Mutagen classifications; *see* page xv for definitions.

Miscellaneous

ACGIH® – American Conference of Governmental Industrial Hygienists

ACGIH® TLVs® – ACGIH® Threshold Limit Values

AIHA – American Industrial Hygiene Association

AIHA WEELs – AIHA Workplace Environmental Exposure Levels

BEI – ACGIH® has recommended a Biological Exposure Index or Indices (BEIs®) for this substance:
BEI$_A$ = Acetylcholinesterase Inhibiting Pesticides;
BEI$_M$ = Methemoglobin Inducers; and
BEI$_P$ = Polycyclic Aromatic Hydrocarbons (PAHs)
all of which are contained in the BEI® section of the current *TLVs®* and *BEIs®* book. For proper application, read the *BEI® Documentation* for the substance.

CAS – Chemical Abstracts Service Registry Number

Ceiling (C) – The concentration that shall not be exceeded during any part of the working exposure

MAK – Federal Republic of Germany Maximum Concentration Values at the Workplace

NIC – Notice of Intended Changes

NIOSH Ceiling – The exposure that shall not be exceeded during any part of the workday. If instantaneous monitoring is not feasible, the ceiling shall be assessed as a 15-minute TWA exposure (unless otherwise specified) that shall not be exceeded at any time during a workday.

NIOSH RELs – U.S. National Institute for Occupational Safety and Health Recommended Exposure Limits. For NIOSH RELs, TWA indicates a time-weighted average concentration for up to a 10-hour workday during a 40-hour workweek.

NIOSH RMLs – NIOSH will no longer use the term REL for occupational carcinogens. Instead, NIOSH will use the term *risk management limit* for carcinogens or RML-CA. An RML-CA is the daily maximum 8-hour time-weighted average concentration of a carcinogen above which a worker should not be exposed. Chemicals determined to be occupational carcinogens will be identified and updated within this guide as established by NIOSH.

OARS – Occupational Alliance for Risk Science

OARS WEELs – OARS Workplace Environmental Exposure Levels

OSHA PELs – U.S. Occupational Safety and Health Administration Permissible Exposure Limits

Skin – Danger of cutaneous absorption

STEL – Short-Term Exposure Limit. Usually a 15-minute time-weighted average (TWA) exposure that should not be exceeded at any time during a workday, even if the 8-hour TWA is within the TLV–TWA, PEL–TWA, or REL–TWA

TERA – Toxicological Excellence for Risk Assessment

TWA – Time-weighted average exposure concentration for a conventional 8-hour (TLV®, PEL) or up to a 10-hour (REL) workday and a 40-hour workweek

( ) – Values/notations contained in parentheses under the ACGIH® TLV® column indicate that these are under review and that an NIC exists.

MAK EXCURSION FACTORS, MAXIMUM DURATION OF PEAKS, MAXIMUM NUMBER PER SHIFT, AND MINIMUM INTERVAL BETWEEN PEAKS

Category	Excursion Factor	Duration	Number per Shift	Interval[A]
I Substances for which local Irritant effects determine the MAK value, also respiratory allergens	1[B]	15 min, average value[C]	4	1 hour
II Substances with systemic effects	2[B]	15 min, average value	4	1 hour

[A] Only for excursion factors > 1.

[B] Default value, or a substance-specific value (maximum 8).

[C] In certain cases, a momentary value (concentration that should not be exceeded at any time) can also be established.

MAK PREGNANCY RISK GROUP CLASSIFICATION

Group A: Damage to the embryo or foetus in humans has been unequivocally demonstrated and is to be expected even when MAK and BAT values are observed.

Group B: According to currently available information, damage to the embryo or foetus must be expected even when MAK and BAT values are observed.

Group C: There is no reason to fear damage to the embryo or foetus when MAK and BAT values are observed.

Group D: Either there are no data for an assessment of damage to the embryo or foetus or the currently available data are not sufficient for classification in one of the groups A–C.

MAK GERM CELL MUTAGENS (confirmed or suspected)*

1. Germ cell mutagens which have been shown to increase the mutant frequency in the progeny of exposed humans.

2. Germ cell mutagens which have been shown to increase the mutant frequency in the progeny of exposed mammals.

3A. Substances which have been shown to induce genetic damage in germ cells of humans or animals, or which produce mutagenic effects in somatic cells of mammals *in vivo* and have been shown to reach the germ cells in an active form.

3B. Substances which are suspected of being germ cell mutagens because of their genotoxic effects in mammalian somatic cells *in vivo*; in exceptional cases, substances for which there are no *in vivo* data but which are clearly mutagenic *in vitro* and structurally related to known *in vivo* mutagens.

4. Not applicable. (Category 4 carcinogenic substances are those with nongenotoxic mechanisms of action. By definition, germ cell mutagens are genotoxic. Therefore, a Category 4 for germ cell mutagens cannot apply. At some time in the future, it is conceivable that a Category 4 could be established for genotoxic substances with primary targets other than DNA [e.g., purely aneugenic substances] if research results make this seem sensible.)

5. Germ cell mutagens or suspected substances (according to the definition of Category 3A and 3B), the potency of which is considered to be so low that, provided the MAK value is observed, their contribution to genetic risk for humans is expected not to be significant.

*The Categories for classification of germ cell mutagens have been established by analogy with the categories for carcinogenic chemicals at the workplace.

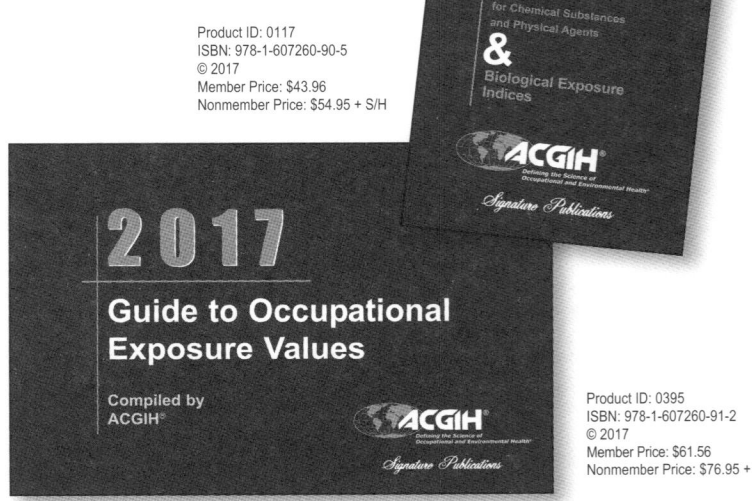

SUBSTANCE / CAS#	ACGIH® TLVs® TWA ppm	mg/m³	STEL/CEIL(C) ppm	mg/m³	OSHA PELs TWA ppm	mg/m³	STEL/CEIL(C) ppm	mg/m³	NIOSH RELs TWA ppm	mg/m³	STEL/CEIL(C) ppm	mg/m³	DFG MAKs TWA ppm	mg/m³	PEAK/CEIL(C) ppm	mg/m³	AIHA WEELs TWA ppm	mg/m³	STEL/CEIL(C) ppm	mg/m³	CARCINOGENICITY CATEGORY
Abietic acid 514-10-3															Sh						
Acenaphthene 83-32-9																					EPA-A* IARC-3 *oral route
Acenaphthylene 208-96-8																					EPA-D
Acephate 30560-19-1																					EPA-C
Acepyrene 25732-74-5																					IARC-3
Acetaldehyde (Acetic aldehyde) 75-07-0			C 25	C 45	200	360			*See* Pocket Guide Apps. A and C				50	91	I (1) C 100 C 180 C; 5						EPA-B2 NTP-R IARC-2B TLV-A2 MAK-5 NIOSH-Ca
Acetamide 60-35-5	1 IFV	2.24 IFV																			IARC-2B MAK-3B TLV-A3
Acetaminophen (Paracetamol) 103-90-2																					IARC-3
Acetic acid 64-19-7	10	25	15	37	10	25			10	25	15	37	10	25	I (2) C						

SUBSTANCE / CAS#	ACGIH® TLVs® TWA ppm	mg/m³	STEL/CEIL(C) ppm	mg/m³	OSHA PELs TWA ppm	mg/m³	STEL/CEIL(C) ppm	mg/m³	NIOSH RELs TWA ppm	mg/m³	STEL/CEIL(C) ppm	mg/m³	DFG MAKs TWA ppm	mg/m³	PEAK/CEIL(C) ppm	mg/m³	AIHA WEELs TWA ppm	mg/m³	STEL/CEIL(C) ppm	mg/m³	CARCINOGENICITY CATEGORY
Acetic anhydride 108-24-7	1	4	C 3		5	20					C 5	C 20	5	21	I (1) D						TLV-A4
Acetone 67-64-1	250	594 BEI	500	1187	1000	2400			250	590			500	1200	I (2) B						EPA-I TLV-A4
Acetone cyanohydrin 75-86-5			C 5* *as CN Skin								C 1* *15-min	C 4*					2 Skin		5		
Acetonitrile 75-05-8	20	34 Skin			40	70			20	34			20	34 Skin; C	II (2)						EPA-CBD; D TLV-A4
Acetophenone 98-86-2	10	49															10				EPA-D
2-Acetylaminofluorene (2-AAF) 53-96-3					*See* 29 CFR 1910.1014				*See* Pocket Guide App. A												NIOSH-Ca NTP-R OSHA-Ca
Acetyl chloride 75-36-5																					EPA-D
Acetylene 74-86-2	*Documentation* withdrawn; *see* Appendix F in *TLVs® and BEIs®* book (D, EX)										C 2500	C 2662									
Acetylsalicylic acid (Aspirin) 50-78-2		5								5											

SUBSTANCE / CAS#	ACGIH® TLVs® TWA ppm	mg/m³	STEL/CEIL(C) ppm	mg/m³	OSHA PELs TWA ppm	mg/m³	STEL/CEIL(C) ppm	mg/m³	NIOSH RELs TWA ppm	mg/m³	STEL/CEIL(C) ppm	mg/m³	DFG MAKs TWA ppm	mg/m³	PEAK/CEIL(C) ppm	mg/m³	AIHA WEELs TWA ppm	mg/m³	STEL/CEIL(C) ppm	mg/m³	CARCINOGENICITY CATEGORY
Aciclovir 59277-89-3																					IARC-3
Acridine Orange 494-38-2																					IARC-3
Acriflavinium chloride 8018-07-3																					IARC-3
Acrolein 107-02-8			C 0.1	C 0.23	0.1	0.25			0.1	0.25	0.3	0.8									EPA-I IARC-3 MAK-3B TLV-A4
		Skin							*See* Pocket Guide App. C												
Acrylamide 79-06-1		0.03 **IFV**				0.3				0.03											EPA-L NTP-R IARC-2A TLV-A3 MAK-2 NIOSH-Ca
		Skin				Skin			Skin *See* Pocket Guide App. A				Skin; Sh; 2								
Acrylic acid 79-10-7	2	5.9							2	6			10	30	I (1)						IARC-3 TLV-A4
		Skin							Skin				C								
Acrylic acid polymer, neutralized, cross-linked 9003-04-7													0.05 **R**		I (1)						MAK-4
													C								
Acrylonitrile (Vinyl cyanide) 107-13-1	2	4.3			2		C 10		1		C 10*										EPA-B1 NTP-R IARC-2B OSHA-Ca MAK-2 TLV-A3 NIOSH-Ca
		Skin			Skin *See* 29 CFR 1910.1045				Skin *15-min *See* Pocket Guide App. A				Skin; Sh							
Actinomycin D 50-76-0																					IARC-3

SUBSTANCE / CAS#	ACGIH® TLVs® TWA ppm	mg/m³	STEL/CEIL(C) ppm	mg/m³	OSHA PELs TWA ppm	mg/m³	STEL/CEIL(C) ppm	mg/m³	NIOSH RELs TWA ppm	mg/m³	STEL/CEIL(C) ppm	mg/m³	DFG MAKs TWA ppm	mg/m³	PEAK/CEIL(C) ppm	mg/m³	AIHA WEELs TWA ppm	mg/m³	STEL/CEIL(C) ppm	mg/m³	CARCINOGENICITY CATEGORY
Adipic acid 124-04-9		5											2 **I**		I (2) C						
Adiponitrile 111-69-3	2	8.8 Skin							4	18											EPA-D
Adriamycin®, Doxorubicin hydrochloride 23214-92-8																					IARC-2A NTP-R
Aflatoxins 1402-68-2														Skin; 3A							IARC-1; 2B* MAK-1 NTP-K *CAS: 6795-23-9
Agaritine 2757-90-6																					IARC-3
Alachlor 15972-60-8		1 **IFV** DSEN																			TLV-A3
Aldicarb 116-06-3		NIC-0.005 **IFV** NIC-A4 NIC-Skin															0.0001			Skin	EPA-D IARC-3
Aldrin 309-00-2		0.05 **IFV** Skin				0.25 Skin				0.25 Skin *See* Pocket Guide App. A				0.25 **I** Skin		II (8)					EPA-B2 IARC-3 NIOSH-Ca TLV-A3

SUBSTANCE / CAS#	ACGIH® TLVs® TWA ppm	mg/m³	STEL/CEIL(C) ppm	mg/m³	OSHA PELs TWA ppm	mg/m³	STEL/CEIL(C) ppm	mg/m³	NIOSH RELs TWA ppm	mg/m³	STEL/CEIL(C) ppm	mg/m³	DFG MAKs TWA ppm	mg/m³	PEAK/CEIL(C) ppm	mg/m³	AIHA WEELs TWA ppm	mg/m³	STEL/CEIL(C) ppm	mg/m³	CARCINOGENICITY CATEGORY
Aliphatic hydrocarbon gases, Alkanes [C₁–C₄]	TLV® withdrawn. Methane, Ethane, Propane, Liquefied petroleum gas (LPG) and Natural gas – *see* Appendix F in *TLVs® and BEIs®* book. Butane and Isobutane – refer to Butane, isomers																				
Alkali persulfates															Sah						
Allyl alcohol (AA)	0.5	1.19			2	5			2	5	4	10									MAK-3B TLV-A4
107-18-6		Skin				Skin				Skin				Skin							
Allyl bromide	0.1	0.5	0.2	1.0																	TLV-A4
106-95-6		Skin																			
Allyl chloride	1	3	2	6	1	3			1	3	2	6									EPA-C IARC-3 MAK-3B TLV-A3
107-05-1		Skin												Skin							
Allyl glycidyl ether (AGE)	1	4.7					C 10	C 45	5	22	10	44									MAK-2 TLV-A4
106-92-3										Skin				Skin; Sh							
Allyl isothiocyanate																				1	IARC-3
57-06-7														Skin; DSEN							
Allyl isovalerate																					IARC-3
2835-39-4																					

SUBSTANCE / CAS#	ACGIH® TLVs® TWA ppm	mg/m³	STEL/CEIL(C) ppm	mg/m³	OSHA PELs TWA ppm	mg/m³	STEL/CEIL(C) ppm	mg/m³	NIOSH RELs TWA ppm	mg/m³	STEL/CEIL(C) ppm	mg/m³	DFG MAKs TWA ppm	mg/m³	PEAK/CEIL(C) ppm	mg/m³	AIHA WEELs TWA ppm	mg/m³	STEL/CEIL(C) ppm	mg/m³	CARCINOGENICITY CATEGORY
Allyl methacrylate 96-05-9	NIC-1	NIC-5.16	NIC-Skin																		
1-(2-Allyloxy)-2-(2,4-dichlorophenyl)ethyl-1H-imidazole 35554-44-0													2 **I** Skin; C		II (2)						
Allyl propyl disulfide 2179-59-1	0.5	3	DSEN		2	12			2	12	3	18	2	12	I (1)						
Aluminum hydroxide 21645-51-2													4 **I** 1.5 **R** D								
Aluminum oxide (α-Alumina) 1344-28-1		TLV® withdrawn; *see* Aluminum, metal and insoluble compounds			15*; 5** *Total dust **Respirable fraction								4 **I** 1.5 **R** D								MAK-2* *Fibrous dust
Aluminum, metal and insoluble compounds 7429-90-5	1 **R**				15*; 5** *Total dust **Respirable fraction				10*; 5** *Total dust **Respirable fraction				4 **I** 1.5 **R** D								IARC-1* TLV-A4 *production
Aluminum, pyro powders and welding fumes, as Al									5												
Aluminum, soluble salts and alkyls, as Al		TLV® withdrawn; *see* Aluminum, metal and insoluble compounds							2												
Amaranth 915-67-3																					IARC-3

SUBSTANCE / CAS#	ACGIH® TLVs® TWA ppm	TWA mg/m³	STEL/CEIL(C) ppm	mg/m³	OSHA PELs TWA ppm	mg/m³	STEL/CEIL(C) ppm	mg/m³	NIOSH RELs TWA ppm	mg/m³	STEL/CEIL(C) ppm	mg/m³	DFG MAKs TWA ppm	mg/m³	PEAK/CEIL(C) ppm	mg/m³	AIHA WEELs TWA ppm	mg/m³	STEL/CEIL(C) ppm	mg/m³	CARCINOGENICITY CATEGORY
5-Aminoacenaphthene 4657-93-6																					IARC-3
2-Aminoanthraquinone 117-79-3																					IARC-3 NTP-R
p-Aminoazobenzene 60-09-3															Sh						IARC-2B
o-Aminoazotoluene 97-56-3														Skin; Sh; 3B							IARC-2B MAK-2 NTP-R
p-Aminobenzoic acid 150-13-0																		5			IARC-3
2-Aminobutanol 96-20-8													1	3.7	II (2)		Skin; D				
1-Amino-2,4-dibromo-anthraquinone 81-49-2																					IARC-2B NTP-R
2-Amino-3,4-dimethyl-imidazo[4,5-f]quinoline (MeIQ) 77094-11-2																					IARC-2B NTP-R
2-Amino-3,8-dimethyl-imidazo[4,5-f]quinoxaline (MeIQx) 77500-04-0																					IARC-2B NTP-R

SUBSTANCE / CAS#	ACGIH® TLVs® TWA ppm	TWA mg/m³	STEL/CEIL(C) ppm	mg/m³	OSHA PELs TWA ppm	mg/m³	STEL/CEIL(C) ppm	mg/m³	NIOSH RELs TWA ppm	mg/m³	STEL/CEIL(C) ppm	mg/m³	DFG MAKs TWA ppm	mg/m³	PEAK/CEIL(C) ppm	mg/m³	AIHA WEELs TWA ppm	mg/m³	STEL/CEIL(C) ppm	mg/m³	CARCINOGENICITY CATEGORY
3-Amino-1,4-dimethyl-5H-pyrido[4,3-b]indole (Trp-P-1) 62450-06-0																					IARC-2B
4-Aminodiphenyl 92-67-1	Skin; L				See 29 CFR 1910.1003				See Pocket Guide App. A				can also occur as vapor and aerosol Skin; 3A								IARC-1 OSHA-Ca MAK-1 TLV-A1 NIOSH-Ca NTP-K
4-Aminodiphenylamine 101-54-2													Skin; Sh								MAK-3B
2-Aminodipyrido[1,2-a: 3′,2′-d]imidazole (Glu-P-2) 67730-10-3																					IARC-2B
2-(2-Aminoethoxy) ethanol 929-06-6													0.2	0.87	I (1)						
													Skin; Sh; C								
6-Amino-2-ethoxy-naphthalene 293733-21-8																					MAK-2
3-Amino-9-ethyl-carbazole 132-32-1																					MAK-3B
1-Amino-2-methyl-anthraquinone 82-28-0																					IARC-3 NTP-R
2-Amino-2-methyl-1-propanol 124-68-5													1	3.7*	II (2)						
													*can also occur as vapor and aerosol Skin; C								

SUBSTANCE	ACGIH® TLVs®				OSHA PELs				NIOSH RELs				DFG MAKs				AIHA WEELs				CARCINOGENICITY CATEGORY
	TWA		STEL/CEIL(C)		TWA		STEL/CEIL(C)		TWA		STEL/CEIL(C)		TWA		PEAK/CEIL(C)		TWA		STEL/CEIL(C)		
CAS#	ppm	mg/m³	ppm	mg/m³	ppm	mg/m³	ppm	mg/m³	ppm	mg/m³	ppm	mg/m³	ppm	mg/m³	ppm	mg/m³	ppm	mg/m³	ppm	mg/m³	
2-Amino-6-methyldipyri-do[1,2-a:3′,2′-d]imida-zole (Glu-P-1) 67730-11-4																					IARC-2B
2-Amino-3-methylimid-azo[4,5-f]quinoline (IQ) 76180-96-6																					IARC-2A NTP-R
2-Amino-1-methyl-6-phe-nylimidazo[4,5-b]pyridine (PhIP) 105650-23-5																					IARC-2B NTP-R
3-Amino-1-methyl-5H-pyrido[4,3-b]indole (Trp-P-2) 62450-07-1																					IARC-2B
2-Amino-3-methyl-9H-pyrido[2,3-b]indole (MeA-α-C) 68006-83-7																					IARC-2B
3-Aminomethyl-3,5,5-trimethyl cyclohexylamine (Isophorone diamine) 2855-13-2														Sh							
2-Amino-5-(5-nitro-2-furyl)-1,3,4-thiadiazole 712-68-5																					IARC-2B
2-Amino-4-nitrophenol 99-57-0																					IARC-3

SUBSTANCE / CAS#	ACGIH® TLVs® TWA ppm	mg/m³	STEL/CEIL(C) ppm	mg/m³	OSHA PELs TWA ppm	mg/m³	STEL/CEIL(C) ppm	mg/m³	NIOSH RELs TWA ppm	mg/m³	STEL/CEIL(C) ppm	mg/m³	DFG MAKs TWA ppm	mg/m³	PEAK/CEIL(C) ppm	mg/m³	AIHA WEELs TWA ppm	mg/m³	STEL/CEIL(C) ppm	mg/m³	CARCINOGENICITY CATEGORY
2-Amino-5-nitrophenol 121-88-0																					IARC-3
2-Amino-5-nitrothiazole 121-66-4																					IARC-3
p-Aminophenol 123-30-8															Sh						
3-Aminophenyl 591-27-5															Sh						
bis(4-Aminophenyl) ether (4,4'-Oxydianline; 4,4'-Diaminodiphenyl) 101-80-4																					IARC-2B MAK-2 NTP-R
N-(3-Aminopropyl)-N dodecylpropane-1,3-diamine 2372-82-9														0.05 I	II (8) C						
2-Aminopyridine 504-29-0	0.5	2			0.5	2			0.5	2											
4-Aminopyridine 504-24-5																					EPA-D
2-Amino-9H-pyrido [2,3-b]indole (A-α-C) 26148-68-5																					IARC-2B

SUBSTANCE / CAS#	ACGIH® TLVs® TWA ppm	TWA mg/m³	STEL/CEIL(C) ppm	mg/m³	OSHA PELs TWA ppm	mg/m³	STEL/CEIL(C) ppm	mg/m³	NIOSH RELs TWA ppm	mg/m³	STEL/CEIL(C) ppm	mg/m³	DFG MAKs TWA ppm	mg/m³	PEAK/CEIL(C) ppm	mg/m³	AIHA WEELs TWA ppm	mg/m³	STEL/CEIL(C) ppm	mg/m³	CARCINOGENICITY CATEGORY
Aminotris(methylene-phosphonic acid) 6419-19-8																		10			
11-Aminoundecanoic acid 2432-99-7																					IARC-3
Amitrole (3-Amino-1,2,4-triazole) 61-82-5		0.2								0.2			*See* Pocket Guide App. A	0.2 I		II (8)	Skin; C				IARC-3 TLV-A3 MAK-4 NIOSH-Ca NTP-R
Ammonia 7664-41-7	25	17	35	24	50	35			25	18	35	27	20	14	I (2)		C				
Ammonium acetate 631-61-8																					EPA-D
Ammonium chloride fume 12125-02-9		10		20						10		20									
Ammonium methacrylate 16325-47-6																					EPA-D
Ammonium perfluorooctanoate 3825-26-1		0.01											Skin								TLV-A3
Ammonium persulfate, as S_2O_8 7727-54-0		0.1															Sah				

SUBSTANCE / CAS#	ACGIH® TLVs® TWA ppm	TWA mg/m³	STEL/CEIL(C) ppm	STEL/CEIL(C) mg/m³	OSHA PELs TWA ppm	TWA mg/m³	STEL/CEIL(C) ppm	STEL/CEIL(C) mg/m³	NIOSH RELs TWA ppm	TWA mg/m³	STEL/CEIL(C) ppm	STEL/CEIL(C) mg/m³	DFG MAKs TWA ppm	TWA mg/m³	PEAK/CEIL(C) ppm	PEAK/CEIL(C) mg/m³	AIHA WEELs TWA ppm	TWA mg/m³	STEL/CEIL(C) ppm	STEL/CEIL(C) mg/m³	CARCINOGENICITY CATEGORY
Ammonium sulfamate 7773-06-0		10			15*; 5** *Total dust **Respirable fraction				10*; 5** *Total dust **Respirable fraction												
Amsacrine 51264-14-3																					IARC-2B
α-Amylase														Sa							
α-Amylcinnamalde-hyde 122-40-7														Sh							
tert-Amyl methyl ether (TAME) 994-05-8	20	84																			
Angelicin plus ultra-violet A radiation 523-50-2																					IARC-3
Aniline 62-53-3	2	7.6			5 and Homologs	19			and Homologs See Pocket Guide App. A				2	7.7	II (2)						EPA-B2 TLV-A3 IARC-3 MAK-4 NIOSH-Ca
	Skin; BEI				Skin								Skin; Sh; C								
Animal hair, epithelia and other materials derived from animals														Sah							
Anisidine, o-isomer 90-04-0	0.1	0.5				0.5				0.5											IARC-2B MAK-2 NIOSH-Ca TLV-A3
	Skin; BEI$_M$				Skin				Skin See Pocket Guide App. A				Skin								

SUBSTANCE CAS#	ACGIH® TLVs® TWA ppm	mg/m³	STEL/CEIL(C) ppm	mg/m³	OSHA PELs TWA ppm	mg/m³	STEL/CEIL(C) ppm	mg/m³	NIOSH RELs TWA ppm	mg/m³	STEL/CEIL(C) ppm	mg/m³	DFG MAKs TWA ppm	mg/m³	PEAK/CEIL(C) ppm	mg/m³	AIHA WEELs TWA ppm	mg/m³	STEL/CEIL(C) ppm	mg/m³	CARCINOGENICITY CATEGORY
Anisidine, p-isomer 104-94-9	0.1	0.5				0.5				0.5											IARC-3 MAK-3B TLV-A4
		Skin; BEI_M				Skin				Skin				Skin							
Anisidine hydrochloride, o-isomer 134-29-2																					NTP-R
Anthanthrene 191-26-4															Skin						IARC-3 MAK-2
Anthracene 120-12-7																					EPA-D IARC-3
Anthranilic acid 118-92-3																					IARC-3
Anthraquinone 84-65-1																					IARC-2B
Antimony [7440-36-0] and compounds, as Sb		0.5				0.5				0.5			and inorganic compounds excluding Stibine 3B								MAK-2
Antimony hydride (Stibine) 7803-52-3	0.1	0.51			0.1	0.5			0.1	0.5											
Antimony oxide 1327-33-9														3B							MAK-2

SUBSTANCE / CAS#	ACGIH® TLVs® TWA ppm	TWA mg/m³	STEL/CEIL(C) ppm	STEL/CEIL(C) mg/m³	OSHA PELs TWA ppm	TWA mg/m³	STEL/CEIL(C) ppm	STEL/CEIL(C) mg/m³	NIOSH RELs TWA ppm	TWA mg/m³	STEL/CEIL(C) ppm	STEL/CEIL(C) mg/m³	DFG MAKs TWA ppm	TWA mg/m³	PEAK/CEIL(C) ppm	PEAK/CEIL(C) mg/m³	AIHA WEELs TWA ppm	TWA mg/m³	STEL/CEIL(C) ppm	STEL/CEIL(C) mg/m³	CARCINOGENICITY CATEGORY
Antimony trioxide, as Sb 1309-64-4	0.5				0.5				0.5						3B						IARC-2B MAK-2
Antimony trioxide 1309-64-4	NIC-0.03 **R** / NIC-A3		(L)																		(TLV-A2)
Antimony trisulfide 1345-04-6																					IARC-3
ANTU (α-Naphthylthiourea) 86-88-4	0.3		Skin		0.3				0.3						Skin						IARC-3 MAK-3B TLV-A4
Apholate 52-46-0																					IARC-3
Apollo 74115-24-5																					EPA-C
p–Aramide, fibrous dust 26125-61-1																					MAK-3B
Aramite® 140-57-8																					EPA-B2 IARC-2B
Argon 7440-37-1	*Documentation* withdrawn; *see* Appendix F in *TLVs® and BEIs®* book Simple asphyxiant (D)																				

SUBSTANCE CAS#	ACGIH® TLVs TWA ppm	mg/m³	STEL/CEIL(C) ppm	mg/m³	OSHA PELs TWA ppm	mg/m³	STEL/CEIL(C) ppm	mg/m³	NIOSH RELs TWA ppm	mg/m³	STEL/CEIL(C) ppm	mg/m³	DFG MAKs TWA ppm	mg/m³	PEAK/CEIL(C) ppm	mg/m³	AIHA WEELs TWA ppm	mg/m³	STEL/CEIL(C) ppm	mg/m³	CARCINOGENICITY CATEGORY
Aristolochic acid 313-67-7																					IARC-1 NTP-K
Arsenic [7440-38-2] and inorganic compounds (except arsine), as As	0.01 BEI				0.5**; 0.01* **Organic cmpds *Inorganic cmpds 29 CFR 1910.1018 Inorgan. cmpds.						C 0.002* *15-min *See* Pocket Guide App. A			Skin; 3A							EPA-A NTP-K IARC-1 OSHA-Ca MAK-1 TLV-A1 NIOSH-Ca
Arsenic acid [7778-39-4] and its salts, as As	0.01														Skin; 3A						EPA-A NTP-K IARC-1 OSHA-Ca MAK-1 TLV-A1 NIOSH-Ca
Arsenic pentoxide, as As 1303-28-2	0.01														Skin; 3A						MAK-1 TLV-A1
Arsenic trioxide, as As 1327-53-3	0.01														Skin; 3A						MAK-1 TLV-A1
Arsenous [13464-58-9] acid and its salts, as As	0.01														Skin; 3A						EPA-A NTP-K IARC-1 OSHA-Ca MAK-1 TLV-A1 NIOSH-Ca
Arsine 7784-42-1	0.005	0.01			0.05	0.2					C 0.002* *15-min *See* Pocket Guide App. A										NIOSH-Ca
Asbestos, all forms 1332-21-4; 12001-28-4; 12172-73-5; 77536-66-4; 77536-67-5; 77536-68-6; 132207-32-0	0.1 f/cc (F)				0.1 f/cc		1 f/cc* *30-min *See* 29 CFR 1910.1001		*See* Pocket Guide Apps. A and C												EPA-A EPA-CaH* IARC-1 MAK-1 NIOSH-Ca NTP-K OSHA-Ca TLV-A1 * Libby Amphibole Asbestos

SUBSTANCE / CAS#	ACGIH® TLVs® TWA ppm	mg/m³	STEL/CEIL(C) ppm	mg/m³	OSHA PELs TWA ppm	mg/m³	STEL/CEIL(C) ppm	mg/m³	NIOSH RELs TWA ppm	mg/m³	STEL/CEIL(C) ppm	mg/m³	DFG MAKs TWA ppm	mg/m³	PEAK/CEIL(C) ppm	mg/m³	AIHA WEELs TWA ppm	mg/m³	STEL/CEIL(C) ppm	mg/m³	CARCINOGENICITY CATEGORY
Asphalt fume (Bitumen) 8052-42-4	0.5 I as benzene-soluble aerosol BEI$_P$										C 5* *15-min *See* Pocket Guide Apps. A and C			Skin; (V)						IARC-2B*; 2A** NIOSH-Ca MAK-2 TLV-A4 * hard bitumens and emissions during mastic asphalt work; straight run bitumens emissions during road paving **oxidized bitumens emissions during roofing
Assure 76578-14-8																					EPA-D
Atrazine 1912-24-9	2 I								5				1 I		II (2) C						IARC-3 TLV-A3
Attapulgite, fibrous dust (Palygorskite) 12174-11-7																					IARC-2B*; 3** MAK-2 *> 5 µm **< 5 µm
Auramine 492-80-8														Skin; 3B							IARC-1*; 2B MAK-2 *production
Auramine hydrochloride 2465-27-2														Skin; 3B							MAK-2
Aurothioglucose 12192-57-3																					IARC-3
Azacitidine 320-67-2																					IARC-2A NTP-R

SUBSTANCE / CAS#	ACGIH® TLVs® TWA ppm	mg/m³	STEL/CEIL(C) ppm	mg/m³	OSHA PELs TWA ppm	mg/m³	STEL/CEIL(C) ppm	mg/m³	NIOSH RELs TWA ppm	mg/m³	STEL/CEIL(C) ppm	mg/m³	DFG MAKs TWA ppm	mg/m³	PEAK/CEIL(C) ppm	mg/m³	AIHA WEELs TWA ppm	mg/m³	STEL/CEIL(C) ppm	mg/m³	CARCINOGENICITY CATEGORY
Azaserine 115-02-6																					IARC-2B
Azathioprine 446-86-6																					IARC-1 NTP-K
Azinphos-methyl 86-50-0	0.2 IFV Skin; DSEN; BEI_A				0.2 Skin				0.2 Skin				0.2 I Skin		II (8)						TLV-A4
tris(Aziridinyl)-p-benzo-quinone (Triaziquone) 68-76-8																					IARC-3
2-(1-Aziridinyl)ethanol 1072-52-2																					IARC-3
bis(1-Aziridinyl)morpho-linophosphine sulfide (Morzid) 2168-68-5																					IARC-3
tris(1-Aziridinyl)-phos-phine oxide 545-55-1																					IARC-3
2,4,6-tris(1-Aziridinyl)-s-triazine 51-18-3																					IARC-3
Aziridyl benzoquinone 800-24-8																					IARC-3

SUBSTANCE / CAS#	ACGIH® TLVs® TWA ppm	mg/m³	STEL/CEIL(C) ppm	mg/m³	OSHA PELs TWA ppm	mg/m³	STEL/CEIL(C) ppm	mg/m³	NIOSH RELs TWA ppm	mg/m³	STEL/CEIL(C) ppm	mg/m³	DFG MAKs TWA ppm	mg/m³	PEAK/CEIL(C) ppm	mg/m³	AIHA WEELs TWA ppm	mg/m³	STEL/CEIL(C) ppm	mg/m³	CARCINOGENICITY CATEGORY
Azobenzene 103-33-3																					EPA-B2 IARC-3
Barium [7440-39-3] and soluble compounds, as Ba		0.5				0.5				0.5				0.5 I	II (8) Soluble compounds only D						EPA-CBD*; NL**; D TLV-A4 *inhalation **oral
Barium sulfate 7727-43-7	5 I	E			15*; 5**		*Total dust **Respirable fraction		10*; 5**		*Total dust **Respirable fraction			4 I 0.3 R* *multiplicated with the material density C	II (8)						MAK-4* *respirable fraction
Bendiocarb 22781-23-3	NIC-0.011* NIC-0.1* *IFV NIC-A4 NIC-Skin																				
Benomyl 17804-35-2	1 I	DSEN			15*; 5** *Total dust **Respirable fraction									Sh; 3A							TLV-A3
Bentazon (Basagran) 25057-89-0																					EPA-NL; E
11H-Benz[bc]aceanthrylene 202-94-8																					IARC-3
Benz[j]aceanthrylene 202-33-5																					IARC-2B
Benz[l]aceanthrylene 211-91-6																					IARC-3

SUBSTANCE / CAS#	ACGIH® TLVs® TWA ppm	mg/m³	STEL/CEIL(C) ppm	mg/m³	OSHA PELs TWA ppm	mg/m³	STEL/CEIL(C) ppm	mg/m³	NIOSH RELs TWA ppm	mg/m³	STEL/CEIL(C) ppm	mg/m³	DFG MAKs TWA ppm	mg/m³	PEAK/CEIL(C) ppm	mg/m³	AIHA WEELs TWA ppm	mg/m³	STEL/CEIL(C) ppm	mg/m³	CARCINOGENICITY CATEGORY
Benz[a]acridine 225-11-6																					IARC-3
Benz[c]acridine 225-51-4																					IARC-3
Benzal chloride (Benzyl dichloride) 98-87-3													Skin								IARC-2A MAK-2
Benzaldehyde 100-52-7													DSEN				2	4			
Benz[a]anthracene 56-55-3	L; BEI_P												Skin; 3A								EPA-B2 TLV-A2 IARC-2B MAK-2 NTP-R
Benzene 71-43-2	0.5 Skin; BEI	1.6	2.5	8	1* *Table Z-2 for exclusions in 29 CFR 1910.1028(d) See 29 CFR 1910.1028	3*	5*	15*	0.1 See Pocket Guide App. A	1			Skin; 3A								EPA-A; K NTP-K IARC-1 OSHA-Ca MAK-1 TLV-A1 NIOSH-Ca
Benzidine 92-87-5	Skin; L				See 29 CFR 1910.1003				See Pocket Guide Apps. A and C				and its salts Skin								EPA-A NTP-K IARC-1* OSHA-Ca MAK-1 TLV-A1 NIOSH-Ca *including dyes metabolized to Benzidine
1,2-Benzisothiazol-3(2H)-one 2634-33-5													Sh								
Benzo[b]chrysene 214-17-5																					IARC-3

SUBSTANCE CAS#	ACGIH® TLVs® TWA ppm	ACGIH® TLVs® TWA mg/m³	ACGIH® TLVs® STEL/CEIL(C) ppm	ACGIH® TLVs® STEL/CEIL(C) mg/m³	OSHA PELs TWA ppm	OSHA PELs TWA mg/m³	OSHA PELs STEL/CEIL(C) ppm	OSHA PELs STEL/CEIL(C) mg/m³	NIOSH RELs TWA ppm	NIOSH RELs TWA mg/m³	NIOSH RELs STEL/CEIL(C) ppm	NIOSH RELs STEL/CEIL(C) mg/m³	DFG MAKs TWA ppm	DFG MAKs TWA mg/m³	DFG MAKs PEAK/CEIL(C) ppm	DFG MAKs PEAK/CEIL(C) mg/m³	AIHA WEELs TWA ppm	AIHA WEELs TWA mg/m³	AIHA WEELs STEL/CEIL(C) ppm	AIHA WEELs STEL/CEIL(C) mg/m³	CARCINOGENICITY CATEGORY
Benzo[g]chrysene 196-78-1																					IARC-3
Benzo[a]fluoranthene 203-33-8																					IARC-3
Benzo[b]fluoranthene 205-99-2		L; BEI_P												Skin; 3B							EPA-B2 TLV-A2 IARC-2B MAK-2 NTP-R
Benzo[ghi]fluoranthene 203-12-3																					IARC-3
Benzo[j]fluoranthene 205-82-3														Skin; 3B							IARC-2B MAK-2 NTP-R
Benzo[k]fluoranthene 207-08-9														Skin; 3B							EPA-B2 IARC-2B MAK-2 NTP-R
Benzo[a]fluorene 238-84-6																					IARC-3
Benzo[b]fluorene 243-17-4																					IARC-3
Benzo[c]fluorene 205-12-9																					IARC-3

SUBSTANCE / CAS#	ACGIH® TLVs® TWA ppm	mg/m³	STEL/CEIL(C) ppm	mg/m³	OSHA PELs TWA ppm	mg/m³	STEL/CEIL(C) ppm	mg/m³	NIOSH RELs TWA ppm	mg/m³	STEL/CEIL(C) ppm	mg/m³	DFG MAKs TWA ppm	mg/m³	PEAK/CEIL(C) ppm	mg/m³	AIHA WEELs TWA ppm	mg/m³	STEL/CEIL(C) ppm	mg/m³	CARCINOGENICITY CATEGORY
Benzofuran 271-89-6																					IARC-2B
Benzoic acid 65-85-0														0.5 R* *can also occur as vapor and aerosol Skin; C	II (4)						EPA-D
Benzoic acid alkali salts														10 I causes pseudoallergic reactions Skin; C	II (2)						
Benzo[b]naphtho-[2,1-d]-thiophene 239-35-0															Skin; 3B						IARC-3 MAK-2
Benzo[ghi]perylene 191-24-2																					EPA-D IARC-3
Benzo[c]phenanthrene 195-19-7																					IARC-2B
Benzophenone 119-61-9																		0.5			IARC-2B
Benzo[a]pyrene 50-32-8		L; BEI_P			0.2 See Coal tar pitch volatiles				0.1 Coal tar pitch volatiles, Cyclo-hexane-extractable fraction See Pocket Guide Apps. A and C					Skin; 2							EPA-CaH NTP-R IARC-1 TLV-A2 MAK-2 NIOSH-Ca
Benzo[e]pyrene 192-97-2																					IARC-3

SUBSTANCE / CAS#	ACGIH® TLVs® TWA ppm	mg/m³	STEL/CEIL(C) ppm	mg/m³	OSHA PELs TWA ppm	mg/m³	STEL/CEIL(C) ppm	mg/m³	NIOSH RELs TWA ppm	mg/m³	STEL/CEIL(C) ppm	mg/m³	DFG MAKs TWA ppm	mg/m³	PEAK/CEIL(C) ppm	mg/m³	AIHA WEELs TWA ppm	mg/m³	STEL/CEIL(C) ppm	mg/m³	CARCINOGENICITY CATEGORY
Benzoquinone dioxime, p-isomer 105-11-3																					IARC-3
Benzotrichloride (Benzyl trichloride) 98-07-7			C 0.1	C 0.8 Skin										Skin							EPA-B2 TLV-A2 IARC-2A MAK-2 NTP-R
Benzoyl chloride 98-88-4			C 0.5	C 2.8															C 5 Skin; DSEN		IARC-2A MAK-3B TLV-A4
Benzoyl peroxide (Dibenzoyl peroxide) 94-36-0		5				5				5				5 I	I (1)						IARC-3 TLV-A4
Benzyl acetate 140-11-4	10	61																			IARC-3 TLV-A4
Benzyl alcohol 100-51-6													5	22*	I (2)		10				
													*can also occur as vapor and aerosol Skin; C								
Benzyl chloride 100-44-7	1	5.2			1	5					C 1*	C 5* *15-min		Skin							EPA-B2 IARC-2A MAK-2 TLV-A3
Benzylhemiformal 14548-60-8														releases Formaldehyde Sh							
Benzyl Violet 4B 1694-09-3																					IARC-2B

SUBSTANCE / CAS#	ACGIH® TLVs® TWA ppm	mg/m³	STEL/CEIL(C) ppm	mg/m³	OSHA PELs TWA ppm	mg/m³	STEL/CEIL(C) ppm	mg/m³	NIOSH RELs TWA ppm	mg/m³	STEL/CEIL(C) ppm	mg/m³	DFG MAKs TWA ppm	mg/m³	PEAK/CEIL(C) ppm	mg/m³	AIHA WEELs TWA ppm	mg/m³	STEL/CEIL(C) ppm	mg/m³	CARCINOGENICITY CATEGORY
Beryllium [7440-41-7] **and compounds, as Be**		0.00005 I Skin*; DSEN*; RSEN** *soluble cmpds **soluble and insoluble cmpds				0.002 *30 min peak per 8-hr shift	C 0.005; 0.025*			See Pocket Guide App. A	C 0.0005			Sah							EPA-B1; L*; NIOSH-Ca CBD** NTP-K IARC-1 TLV-A1 MAK-1 *inhaled **ingested
Biphenyl (Diphenyl) 92-52-4	0.2	1.3			0.2	1			0.2	1				Skin							EPA-S MAK-3B
Bismuth telluride, Undoped 1304-82-1		10				15*; 5** *Total dust **Respirable fraction				10*; 5** *Total dust **Respirable fraction											TLV-A4
Bismuth telluride, Se-doped, as Bi₂Te₃ 1304-82-1		5								5											TLV-A4
Bisphenol A (4,4′-Isopropylidenediphenol; BPA) 80-05-7														5 I SP; C		I (1)					
Bisphenol A diglycidyl-ether (4,4′-Isopropylidenediphenol diglycidyl ether) 1675-54-3														Skin; Sh							IARC-3 MAK-3A
Bisphenol A digly-cidyl methacrylate 1565-94-2														Sh							
Bisphenol A glycerolate 4687-94-9														Sh							
Bisphenyl A ethoxylate dimethylacrylate 24448-20-2														Sh							

SUBSTANCE / CAS#	ACGIH® TLVs® TWA ppm	ACGIH® TLVs® TWA mg/m³	STEL/CEIL(C) ppm	STEL/CEIL(C) mg/m³	OSHA PELs TWA ppm	OSHA PELs TWA mg/m³	STEL/CEIL(C) ppm	STEL/CEIL(C) mg/m³	NIOSH RELs TWA ppm	NIOSH RELs TWA mg/m³	STEL/CEIL(C) ppm	STEL/CEIL(C) mg/m³	DFG MAKs TWA ppm	DFG MAKs TWA mg/m³	PEAK/CEIL(C) ppm	PEAK/CEIL(C) mg/m³	AIHA WEELs TWA ppm	AIHA WEELs TWA mg/m³	STEL/CEIL(C) ppm	STEL/CEIL(C) mg/m³	CARCINOGENICITY CATEGORY
Bisphenyl F diglycidyl ether (o,o'-, o,p'-, p,p'-isomers) 54208-63-8; 57469-07-5; 2095-03-6														Sh							
Bisulfites																					IARC-3
Bithionol 97-18-7														SP							
Bleomycins 11056-06-7																					IARC-2B
Blue VRS 129-17-9																					IARC-3
Borate compounds, inorganic 1303-96-4; 1330-43-4; 10043-35-3; 12179-04-3	2 I		6 I						*See* Sodium tetraborate, anhydrous; Sodium tetraborate, decahydrate; and Sodium tetraborate, pentahydrate					10 I* 0.75 I** B*; C**	I (1)* I (1)**						TLV-A4
														*CAS: 10043-35-3 only **Tetraborates, as B							
Boron [7440-42-8] **and compounds**																					EPA-I
Boron oxide 1303-86-2		10				15* *Total dust				10											

SUBSTANCE	ACGIH® TLVs®				OSHA PELs				NIOSH RELs				DFG MAKs				AIHA WEELs				CARCINOGENICITY
	TWA		STEL/CEIL(C)		TWA		STEL/CEIL(C)		TWA		STEL/CEIL(C)		TWA		PEAK/CEIL(C)		TWA		STEL/CEIL(C)		
CAS#	ppm	mg/m³	ppm	mg/m³	ppm	mg/m³	ppm	mg/m³	ppm	mg/m³	ppm	mg/m³	ppm	mg/m³	ppm	mg/m³	ppm	mg/m³	ppm	mg/m³	CATEGORY
Boron tribromide 10294-33-4			C 0.7	C 7							C 1	C 10									
Boron trichloride 10294-34-5			C 0.7	C 2.4																	
Boron trifluoride 7637-07-2	0.1	0.28	C 0.7	C 1.95			C 1	C 3			C 1	C 3									
Boron trifluoride ethers 109-63-7; 353-42-4	NIC-0.1*		NIC-C 0.7*																		
			*as BF₃																		
Brilliant Blue FCF, disodium salt 3844-45-9																					IARC-3
Bromacil 314-40-9 .		10							1	10											TLV-A3
Bromate 15541-45-4																					EPA-L*; I**; B1 *oral route **inhalation
Bromelain 9001-00-7															Sa						
Brominated dibenzo-furans																					EPA-D

SUBSTANCE CAS#	ACGIH® TLVs® TWA ppm	ACGIH® TLVs® TWA mg/m³	ACGIH® TLVs® STEL/CEIL(C) ppm	ACGIH® TLVs® STEL/CEIL(C) mg/m³	OSHA PELs TWA ppm	OSHA PELs TWA mg/m³	OSHA PELs STEL/CEIL(C) ppm	OSHA PELs STEL/CEIL(C) mg/m³	NIOSH RELs TWA ppm	NIOSH RELs TWA mg/m³	NIOSH RELs STEL/CEIL(C) ppm	NIOSH RELs STEL/CEIL(C) mg/m³	DFG MAKs TWA ppm	DFG MAKs TWA mg/m³	DFG MAKs PEAK/CEIL(C) ppm	DFG MAKs PEAK/CEIL(C) mg/m³	AIHA WEELs TWA ppm	AIHA WEELs TWA mg/m³	AIHA WEELs STEL/CEIL(C) ppm	AIHA WEELs STEL/CEIL(C) mg/m³	CARCINOGENICITY CATEGORY
Bromine 7726-95-6	0.1	0.66	0.2	1.3	0.1	0.7			0.1	0.7	0.3	2									
Bromine pentafluoride 7789-30-2	0.1	0.72							0.1	0.7											
Bromobenzene 108-86-1																					EPA-II
Bromochloroacetic acid 5589-96-8																					IARC-2B
Bromochloroace-tonitrile 83463-62-1																					IARC-3
Bromodichloro-methane 75-27-4														Skin; 3B							EPA-B2 IARC-2B MAK-2 NTP-R
p-Bromodiphenyl ether 101-55-3																					EPA-D
Bromoform (Tribromomethane) 75-25-2	0.5	5.2			0.5	5		Skin	0.5	5		Skin									EPA-B2 IARC-3 MAK-3B TLV-A3
2,2-bis-(Bromomethyl)-1,3-propanediol, technical grade 3296-90-0																		0.2			IARC-2B NTP-R

SUBSTANCE / CAS#	ACGIH® TLVs® TWA ppm	mg/m³	STEL/CEIL(C) ppm	mg/m³	OSHA PELs TWA ppm	mg/m³	STEL/CEIL(C) ppm	mg/m³	NIOSH RELs TWA ppm	mg/m³	STEL/CEIL(C) ppm	mg/m³	DFG MAKs TWA ppm	mg/m³	PEAK/CEIL(C) ppm	mg/m³	AIHA WEELs TWA ppm	mg/m³	STEL/CEIL(C) ppm	mg/m³	CARCINOGENICITY CATEGORY
2-Bromo-2-nitro-1,3-propanediol 52-51-7														Skin; Sh							
1-Bromopropane 106-94-5	0.1	0.5												Skin							MAK-2 NTP-R TLV-A3
Bromotrichloro-methane 75-62-7																					EPA-D
1,3-Butadiene 106-99-0	2	4.4			1	2.21	5	11						2							EPA-CaH* NIOSH-Ca IARC-1 NTP-K MAK-1 TLV-A2 *inhaled
					See 29 CFR 1910.1051; 29 CFR 1910.19(l)				*See* Pocket Guide App. A												
Butane, isomers 75-28-5; 106-97-8			1000* *(EX)	2370*					800	1900			1000	2400	II (4)						
									CAS: 106-97-8 only					D							
1,4-Butanediol diacrylate 1070-70-8														Sh							
1,4-Butanediol diglycidyl ether 2425-79-8														Sh							
1,4-Butanediol dimethacrylate 2082-81-7														Sh							
1,4-Butanediol dimethanesulfonate (Busulphan) 55-98-1																					IARC-1 NTP-K

SUBSTANCE / CAS#	ACGIH® TLVs® TWA ppm	TWA mg/m³	STEL/CEIL(C) ppm	STEL/CEIL(C) mg/m³	OSHA PELs TWA ppm	TWA mg/m³	STEL/CEIL(C) ppm	STEL/CEIL(C) mg/m³	NIOSH RELs TWA ppm	TWA mg/m³	STEL/CEIL(C) ppm	STEL/CEIL(C) mg/m³	DFG MAKs TWA ppm	TWA mg/m³	PEAK/CEIL(C) ppm	PEAK/CEIL(C) mg/m³	AIHA WEELs TWA ppm	TWA mg/m³	STEL/CEIL(C) ppm	STEL/CEIL(C) mg/m³	CARCINOGENICITY CATEGORY
1,4-Butane sultone 1633-83-6																					MAK-3B
2,4-Butane sultone 1121-03-5																					MAK-2
n-Butanol (n-Butyl alcohol) 71-36-3	20	61			100	300					C 50	C 150 Skin	100	310	I (1)	C					EPA-D
sec-Butanol (sec-Butyl alcohol) 78-92-2	100	300			150	450			100	305	150	455									
tert-Butanol (tert-Butyl alcohol) 75-65-0	100	303			100	300			100	300	150	450	20	62	II (4)	C					TLV-A4
Butenes, all isomers 106-98-9; 107-01-7; 115-11-7; 590-18-1; 624-64-6; 25167-67-3	250	574																			TLV-A4* *CAS: 115-11-7 only
2-Butoxyethanol (EGBE) 111-76-2	20	97 BEI			50	240 Skin			5	24 Skin			10*	49	I (2) *sum of the concentrations of EGBE and its acetate in air Skin; C						EPA-NL IARC-3 MAK-4 TLV-A3
2-Butoxyethyl acetate (EGBEA) 112-07-2	20	130							5	33			10*	66	I (2) *sum of the concentrations of EGBE and its acetate in air Skin; C						MAK-4 TLV-A3
1-tert-Butoxy-2-propanol 57018-52-7															C						IARC-3

SUBSTANCE / CAS#	ACGIH® TLVs® TWA ppm	TWA mg/m³	STEL/CEIL(C) ppm	STEL/CEIL(C) mg/m³	OSHA PELs TWA ppm	TWA mg/m³	STEL/CEIL(C) ppm	STEL/CEIL(C) mg/m³	NIOSH RELs TWA ppm	TWA mg/m³	STEL/CEIL(C) ppm	STEL/CEIL(C) mg/m³	DFG MAKs TWA ppm	TWA mg/m³	PEAK/CEIL(C) ppm	PEAK/CEIL(C) mg/m³	AIHA WEELs TWA ppm	TWA mg/m³	STEL/CEIL(C) ppm	STEL/CEIL(C) mg/m³	CARCINOGENICITY CATEGORY
n-Butyl acetate 123-86-4	*Adopted TLV® and Documentation withdrawn; see Butyl acetates, all isomers*				150	710			150	710	200	950	100	480	I (2) C						
sec-Butyl acetate 105-46-4	*Adopted TLV® and Documentation withdrawn; see Butyl acetates, all isomers*				200	950			200	950											
tert-Butyl acetate 540-88-5	*Adopted TLV® and Documentation withdrawn; see Butyl acetates, all isomers*				200	950			200	950			50	240	II (2) C						
Butyl acetates, all isomers 105-46-4; 110-19-0; 123-86-4; 540-88-5	50	238	150	712																	
n-Butyl acrylate (Acrylic acid ester, n-Butyl ester) 141-32-2	2	11		DSEN					10	55			2	11	I (2) Skin; Sh; C						IARC-3 TLV-A4
tert-Butyl acrylate 1663-39-4															Sh						
n-Butylamine 109-73-9			C 5 Skin	C 15			C 5 Skin	C 15			C 5 Skin	C 15	2	6.1	I (2) C 5 C	15					
sec-Butylamine 13952-84-6													2	6.1	I (2) C 5 D	15					
tert-Butyl hydroperoxide 75-91-2	NIC-0.1	NIC-0.4		NIC-Skin																	

SUBSTANCE / CAS#	ACGIH TLVs TWA ppm	TWA mg/m³	STEL/CEIL(C) ppm	STEL/CEIL(C) mg/m³	OSHA PELs TWA ppm	TWA mg/m³	STEL/CEIL(C) ppm	STEL/CEIL(C) mg/m³	NIOSH RELs TWA ppm	TWA mg/m³	STEL/CEIL(C) ppm	STEL/CEIL(C) mg/m³	DFG MAKs TWA ppm	TWA mg/m³	PEAK/CEIL(C) ppm	PEAK/CEIL(C) mg/m³	AIHA WEELs TWA ppm	TWA mg/m³	STEL/CEIL(C) ppm	STEL/CEIL(C) mg/m³	CARCINOGENICITY CATEGORY
Butylated hydroxyanisole (BHA) 25013-16-5														20 I	II (1)						IARC-2B MAK-3B NTP-R
														C							
Butylated hydroxytoluene (BHT; 2,6-Di-tert-butyl-p-cresol) 128-37-0		2 IFV								10				10 I	II (4)						IARC-3 MAK-4 TLV-A4
														C							
4-tert-Butylbenzoic acid 98-73-7														2 I	II (2)						
													Skin; D							
Butyl benzyl phthalate 85-68-7																					EPA-C IARC-3
Butyl carbitol acetate (Diethylene glycol monobutyl ether acetate) 124-17-4													10	85	I (1.5)						
													C							
p-tert-Butylcatechol (4-[1,1-Dimethylethyl]-1,2-benzenediol) 98-29-3																			C 2		
													Sh			Skin; DSEN				
tert-Butylchloride 507-20-0																					EPA-D
n-Butyl chloroformate (Chloroformic acid butyl ester) 592-34-7													0.2	1.1	I (2)						
													C							
tert-Butyl chromate, as CrO₃ 1189-85-1			C 0.1		0.005				0.001*												NIOSH-Ca
		Skin			Skin, See 29 CFR 1910.1026				*as Cr(VI), See Pocket Guide Apps. A and C											

SUBSTANCE / CAS#	ACGIH® TLVs® TWA ppm	mg/m³	STEL/CEIL(C) ppm	mg/m³	OSHA PELs TWA ppm	mg/m³	STEL/CEIL(C) ppm	mg/m³	NIOSH RELs TWA ppm	mg/m³	STEL/CEIL(C) ppm	mg/m³	DFG MAKs TWA ppm	mg/m³	PEAK/CEIL(C) ppm	mg/m³	AIHA WEELs TWA ppm	mg/m³	STEL/CEIL(C) ppm	mg/m³	CARCINOGENICITY CATEGORY
n-Butyl glycidyl ether (BGE) 2426-08-6	3	16		Skin; DSEN	50	270					C 5.6*	C 30* *15-min		Skin; Sh; 2							MAK-3B
tert-Butyl glycidyl ether 7665-72-7														Skin; Sh							MAK-3B
n-Butyl lactate 138-22-7	5	30							5	25											
n-Butyl mercaptan (Butanethiol) 109-79-5	0.5	1.8			10	35					C 0.5*	C 1.8* *15-min	0.5	1.9	II (2) C						
n-Butyl methacrylate 97-88-1														Sh							
o-sec-Butylphenol 89-72-5	5	31		Skin					5	30		Skin									
p-tert-Butylphenol 98-54-4													0.08	0.5	II (2) Skin; Sh; D						
p-tert-Butyl phenyl glycidyl ether 3101-60-8														Sh							
p-tert-Butyltoluene 98-51-1	1	6.1			10	60			10	60	20	120									

SUBSTANCE / CAS#	ACGIH® TLVs® TWA ppm	TWA mg/m³	STEL/CEIL(C) ppm	STEL/CEIL(C) mg/m³	OSHA PELs TWA ppm	TWA mg/m³	STEL/CEIL(C) ppm	STEL/CEIL(C) mg/m³	NIOSH RELs TWA ppm	TWA mg/m³	STEL/CEIL(C) ppm	STEL/CEIL(C) mg/m³	DFG MAKs TWA ppm	TWA mg/m³	PEAK/CEIL(C) ppm	PEAK/CEIL(C) mg/m³	AIHA WEELs TWA ppm	TWA mg/m³	STEL/CEIL(C) ppm	STEL/CEIL(C) mg/m³	CARCINOGENICITY CATEGORY
iso-Butyl vinyl ether 109-53-5													20	83	I (1)						
														D							
Butynediol 110-65-6													0.1	0.36	I (1)						
													Skin; Sh; C							
Butyraldehyde 123-72-8																	25*				
																*OARS WEEL				
β-Butyrolactone 3068-88-0																					IARC-2B
γ-Butyrolactone 96-48-0																					IARC-3
													Skin							
n-Butyronitrile 109-74-0									8	22											
Cacodylic acid 75-60-5																					EPA-D IARC-2B
Cadmium [7440-43-9] and compounds, as Cd		0.01 0.002 **R** BEI			0.005* *Table Z-2 for exclusions in 29 CFR 1910.1027 See 29 CFR 1910.1027				*See* Pocket Guide App. A												EPA-B1 OSHA-Ca IARC-1 TLV-A2 NIOSH-Ca NTP-K
Cadmium [7440-43-9] and inorganic compounds													Skin; 3A								MAK-1

SUBSTANCE CAS#	ACGIH® TLVs® TWA ppm	ACGIH® TLVs® TWA mg/m³	ACGIH® TLVs® STEL/CEIL(C) ppm	ACGIH® TLVs® STEL/CEIL(C) mg/m³	OSHA PELs TWA ppm	OSHA PELs TWA mg/m³	OSHA PELs STEL/CEIL(C) ppm	OSHA PELs STEL/CEIL(C) mg/m³	NIOSH RELs TWA ppm	NIOSH RELs TWA mg/m³	NIOSH RELs STEL/CEIL(C) ppm	NIOSH RELs STEL/CEIL(C) mg/m³	DFG MAKs TWA ppm	DFG MAKs TWA mg/m³	DFG MAKs PEAK/CEIL(C) ppm	DFG MAKs PEAK/CEIL(C) mg/m³	AIHA WEELs TWA ppm	AIHA WEELs TWA mg/m³	AIHA WEELs STEL/CEIL(C) ppm	AIHA WEELs STEL/CEIL(C) mg/m³	CARCINOGENICITY CATEGORY
Cadusafos 95465-99-9	0.00009* *IFV	0.001* Skin																			TLV-A4
Caffeic acid 331-39-5																					IARC-2B
Caffeine 58-08-2																					IARC-3
Calcium arsenate, as As 7778-44-1											C 0.002* *15-min *See* Pocket Guide App. A		*See* Arsenic and inorganic compounds								NIOSH-Ca
Calcium carbonate (Limestone; Marble) 1317-65-3	TLV® withdrawn due to	insufficient data			15*; 5** *Total dust	**Respirable fraction			10*; 5** Includes CAS: 471-34-1 *Total dust	**Respirable fraction											
Calcium chromate, as Cr 13756-19-0	(0.001) NIC-withdraw adopted TLV® and *Documentation*; *see* Chromium and	inorganic compounds							*See* Chromic acid and chromates				*See* Chromium(VI) inorganic compounds, water-soluble								(TLV-A2)
Calcium cyanamide 156-62-7	0.5								0.5				1 I Skin; C		II (2)						TLV-A4
Calcium cyanide, as CN 592-01-8	*See* Hydrogen cyanide and Cyanide salts, as CN				*See* Cyanides				*See* Sodium cyanide				*See* Cyanides								
Calcium hydroxide 1305-62-0	5				15*; 5** *Total dust	**Respirable fraction			5				1 I C		I (2)						

SUBSTANCE / CAS#	ACGIH® TLVs® TWA ppm	TWA mg/m³	STEL/CEIL(C) ppm	STEL/CEIL(C) mg/m³	OSHA PELs TWA ppm	TWA mg/m³	STEL/CEIL(C) ppm	STEL/CEIL(C) mg/m³	NIOSH RELs TWA ppm	TWA mg/m³	STEL/CEIL(C) ppm	STEL/CEIL(C) mg/m³	DFG MAKs TWA ppm	TWA mg/m³	PEAK/CEIL(C) ppm	PEAK/CEIL(C) mg/m³	AIHA WEELs TWA ppm	TWA mg/m³	STEL/CEIL(C) ppm	STEL/CEIL(C) mg/m³	CARCINOGENICITY CATEGORY
Calcium oxide 1305-78-8		2				5				2				1 I		I (2) C					
Calcium silicate synthetic nonfibrous Adopted TLV® and *Documentation* withdrawn; *see* Appendix B in *TLVs*® and *BEIs*® book 1344-95-2						15*; 5**				10*; 5**											TLV-A4
					*Total dust **Respirable fraction				*Total dust **Respirable fraction												
Calcium silicate, naturally occurring as wollastonite		1 I* *nonfibrous E																			TLV-A4
Calcium sodium meta-phosphate (fibrous dust) 23209-59-8																					MAK-3B
Calcium sulfate 7778-18-9; 10034-76-1; 10101-41-4; 13397-24-5		10 I				15*; 5** CAS: 7778-18-9 *Total dust **Respirable fraction				10*; 5** *Total dust **Respirable fraction				4 I 1.5 R C							
Camphor, synthetic 76-22-2	2	12	3	19	2				2												TLV-A4
Cantharidin 56-25-7																					IARC-3
Caprolactam 105-60-2		5 IFV							0.22*	1**	0.66*	3** *vapor **dust		5 I		I (2) C					IARC-4 TLV-A5

SUBSTANCE / CAS#	ACGIH® TLVs® TWA ppm	mg/m³	STEL/CEIL(C) ppm	mg/m³	OSHA PELs TWA ppm	mg/m³	STEL/CEIL(C) ppm	mg/m³	NIOSH RELs TWA ppm	mg/m³	STEL/CEIL(C) ppm	mg/m³	DFG MAKs TWA ppm	mg/m³	PEAK/CEIL(C) ppm	mg/m³	AIHA WEELs TWA ppm	mg/m³	STEL/CEIL(C) ppm	mg/m³	CARCINOGENICITY CATEGORY
Captafol 2425-06-1	0.007* *IFV Skin; DSEN; RSEN	0.1*								0.1 Skin *See* Pocket Guide App. A										IARC-2A NIOSH-Ca NTP-R TLV-A3
Captan 133-06-2		5 I DSEN								5 *See* Pocket Guide App. A										IARC-3 NIOSH-Ca TLV-A3
Carbaryl 63-25-2	0.5 IFV BEI_A; Skin				5					5				5 I Skin	II (4)					IARC-3 TLV-A4
Carbazole 86-74-8																					IARC-2B
Carbendazim 10605-21-7														10 I 5; B	II (4)					
3-Carbethoxypsoralen 20073-24-9																					IARC-3
Carbofuran 1563-66-2	0.1 IFV BEI_A									0.1											TLV-A4
Carbon black 1333-86-4		3 I				3.5				3.5* *0.1 in presence of PAHs *See* Pocket Guide Apps. A and C				as Inhalable dust							IARC-2B TLV-A3 MAK-3B NIOSH-Ca* *in presence of PAHs
Carbon dioxide 124-38-9	5000	9000	30,000	54,000	5000	9000			5000	9000	30,000	54,000	5000	9100	II (2)						

SUBSTANCE / CAS#	ACGIH® TLVs® TWA ppm	TWA mg/m³	STEL/CEIL(C) ppm	STEL/CEIL(C) mg/m³	OSHA PELs TWA ppm	TWA mg/m³	STEL/CEIL(C) ppm	STEL/CEIL(C) mg/m³	NIOSH RELs TWA ppm	TWA mg/m³	STEL/CEIL(C) ppm	STEL/CEIL(C) mg/m³	DFG MAKs TWA ppm	TWA mg/m³	PEAK/CEIL(C) ppm	PEAK/CEIL(C) mg/m³	AIHA WEELs TWA ppm	TWA mg/m³	STEL/CEIL(C) ppm	STEL/CEIL(C) mg/m³	CARCINOGENICITY CATEGORY
Carbon disulfide 75-15-0	1	3.13		Skin; BEI	20		C 30; 100* *30-min peak per 8-hr shift		1	3	10	30 Skin	5	16	II (2) Skin; B						TLV-A4
Carbon monoxide 630-08-0	25	29		BEI	50	55			35	40	C 200	C 229	30	35	II (2) B						
Carbon tetrabromide 558-13-4	0.1	1.4	0.3	4.1					0.1	1.4	0.3	4									
Carbon tetrachloride (Tetrachloromethane) 56-23-5	5	31	10	63 Skin	10		C 25; 200* *5-min peak in any 4 hrs				2* *60-min See Pocket Guide App. A	12.6*	0.5	3.2	II (2) Skin; C						EPA-L NTP-R IARC-2B TLV-A2 MAK-4 NIOSH-Ca
Carbonyl fluoride 353-50-4	2	5.4	5	13					2	5	5	15									
Carbonyl sulfide 463-58-1	5	12																			
N-Carboxyanthranilic anhydride 118-48-9															Sh						
Carfentrazone-ethyl 128639-02-1	NIC-1 I			NIC-A4																	
Carmoisine 3567-69-9																					IARC-3

SUBSTANCE / CAS#	ACGIH® TLVs® TWA ppm	TWA mg/m³	STEL/CEIL(C) ppm	STEL/CEIL(C) mg/m³	OSHA PELs TWA ppm	TWA mg/m³	STEL/CEIL(C) ppm	STEL/CEIL(C) mg/m³	NIOSH RELs TWA ppm	TWA mg/m³	STEL/CEIL(C) ppm	STEL/CEIL(C) mg/m³	DFG MAKs TWA ppm	TWA mg/m³	PEAK/CEIL(C) ppm	PEAK/CEIL(C) mg/m³	AIHA WEELs TWA ppm	TWA mg/m³	STEL/CEIL(C) ppm	STEL/CEIL(C) mg/m³	CARCINOGENICITY CATEGORY
Carrageenan, degraded 53973-98-1																					IARC-2B
Carrageenan, native 9000-07-1																					IARC-3
Catechol (Pyrocatechol) 120-80-9	5	23		Skin					5	20		Skin									IARC-2B TLV-A3
Cellulases															Sa						
Cellulose 9004-34-6		10				15*; 5**				10*; 5**											
					*Total dust **Respirable fraction				*Total dust **Respirable fraction												
Cereal flour dusts (Rye, wheat)															Sa						
Cerium oxide and cerium compounds 1306-38-3																					EPA-I
Cesium hydroxide 21351-79-1		2								2											
Cetylmercaptan (1-Hexadecanethiol) 2917-26-2									C 0.5*	C 5.3*											
									*15-min												

SUBSTANCE / CAS#	ACGIH® TLVs® TWA ppm	TWA mg/m³	STEL/CEIL(C) ppm	STEL/CEIL(C) mg/m³	OSHA PELs TWA ppm	TWA mg/m³	STEL/CEIL(C) ppm	STEL/CEIL(C) mg/m³	NIOSH RELs TWA ppm	TWA mg/m³	STEL/CEIL(C) ppm	STEL/CEIL(C) mg/m³	DFG MAKs TWA ppm	TWA mg/m³	PEAK/CEIL(C) ppm	PEAK/CEIL(C) mg/m³	AIHA WEELs TWA ppm	TWA mg/m³	STEL/CEIL(C) ppm	STEL/CEIL(C) mg/m³	CARCINOGENICITY CATEGORY
Chimney sweeping																					IARC-1
Chloral 75-87-6																					IARC-2A
Chloral hydrate 302-17-0																					EPA-C; CBD* IARC-2A *oral
Chlorambucil 305-03-3																					IARC-1 NTP-K
Chloramine 10599-90-3																					EPA-D IARC-3
Chloramphenicol 56-75-7																		0.5			IARC-2A NTP-R
Chlordane 57-74-9		0.5				0.5				0.5 Skin See Pocket Guide App. A				0.5 I Skin	II (8)						EPA-L*; B2 NIOSH-Ca IARC-2B TLV-A3 MAK-3B *CAS: 12789-03-6
	Skin				Skin																
Chlordecone 143-50-0										0.001 See Pocket Guide App. A				Skin							EPA-L* NIOSH-Ca IARC-2B NTP-R MAK-2 *oral
Chlordimeform 6164-98-3																					IARC-3

SUBSTANCE / CAS#	ACGIH® TLVs® TWA ppm	TWA mg/m³	STEL/CEIL(C) ppm	STEL/CEIL(C) mg/m³	OSHA PELs TWA ppm	TWA mg/m³	STEL/CEIL(C) ppm	STEL/CEIL(C) mg/m³	NIOSH RELs TWA ppm	TWA mg/m³	STEL/CEIL(C) ppm	STEL/CEIL(C) mg/m³	DFG MAKs TWA ppm	TWA mg/m³	PEAK/CEIL(C) ppm	PEAK/CEIL(C) mg/m³	AIHA WEELs TWA ppm	TWA mg/m³	STEL/CEIL(C) ppm	STEL/CEIL(C) mg/m³	CARCINOGENICITY CATEGORY
Chlorendic acid 115-28-6																					IARC-2B NTP-R
Chlorinated biphenyls: higher chlorinated (≥ 4 Cl) biphenyls														0.003 I* *(PCB 28 + PCB 52 + PCB 101 + PCB 138 + PCB 153 + PCB 180) × 5 Skin; B; 5	II (8)						MAK-4
Chlorinated biphenyls: mono-, di-, trichlorinated biphenyls														0.003 I Skin; B; 5	II (8)						MAK-4
Chlorinated camphene (Toxaphene) 8001-35-2	0.5		1		0.5									Skin							EPA-B2 NTP-R IARC-2B TLV-A3 MAK-2 NIOSH-Ca
		Skin				Skin			Skin See Pocket Guide App. A												
o-Chlorinated diphenyl oxide 31242-93-0	0.5				0.5				0.5					several CAS Nos., e.g., 55720-99-5 Skin							
Chlorinated drinking water																					IARC-3
Chlorinated naphthalenes														Skin							
Chlorinated paraffins, 20%–70% chlorine																					IARC-2B* MAK-3B NTP-R* *60%

SUBSTANCE / CAS#	ACGIH® TLVs® TWA ppm	mg/m³	STEL/CEIL(C) ppm	mg/m³	OSHA PELs TWA ppm	mg/m³	STEL/CEIL(C) ppm	mg/m³	NIOSH RELs TWA ppm	mg/m³	STEL/CEIL(C) ppm	mg/m³	DFG MAKs TWA ppm	mg/m³	PEAK/CEIL(C) ppm	mg/m³	AIHA WEELs TWA ppm	mg/m³	STEL/CEIL(C) ppm	mg/m³	CARCINOGENICITY CATEGORY
α-Chlorinated toluenes, mixture of benzoyl chloride, benzyl chloride, benzyl dichloride and benzyl trichloride														Skin							IARC-2A MAK-1
Chlorine 7782-50-5	(0.5) NIC-0.1	(1.5) NIC-0.29	(1) NIC-0.4	(2.9) NIC-1.16			C 1	C 3			C 0.5* *15-min	C 1.45*	0.5	1.5	I (1) C						TLV-A4
Chlorine dioxide 10049-04-4	(0.1)	(0.28)	(0.3) NIC-C 0.1	(0.83) NIC-C 0.28	0.1	0.3			0.1	0.3	0.3	0.9	0.1	0.28	I (1) D						EPA-CBD; D
Chlorine trifluoride 7790-91-2			C 0.1	C 0.38			C 0.1	C 0.4			C 0.1	C 0.4									
Chlorite, sodium salt 7758-19-2																					EPA-CBD; D IARC-3
Chloroacetaldehyde 107-20-0			C 1	C 3.2			C 1	C 3			C 1	C 3		Skin							MAK-3B
2-Chloroacetamide 79-07-2														Skin; Sh							
Chloroacetamide-N-methylol (CAM) 2832-19-1														releases Formaldehyde Sh							MAK-3B

SUBSTANCE / CAS#	ACGIH® TLVs® TWA ppm	mg/m³	STEL/CEIL(C) ppm	mg/m³	OSHA PELs TWA ppm	mg/m³	STEL/CEIL(C) ppm	mg/m³	NIOSH RELs TWA ppm	mg/m³	STEL/CEIL(C) ppm	mg/m³	DFG MAKs TWA ppm	mg/m³	PEAK/CEIL(C) ppm	mg/m³	AIHA WEELs TWA ppm	mg/m³	STEL/CEIL(C) ppm	mg/m³	CARCINOGENICITY CATEGORY
Chloroacetic acid, methyl ester (Methyl chloroacetate) 96-34-4													1	4.5	I (1)						
													Skin; Sh; C								
Chloroacetone 78-95-5			C 1	C 3.8																	
			Skin																		
Chloroacetonitrile 107-14-2																					IARC-3
2-Chloroacetophenone (Phenacyl chloride) 532-27-4	0.05	0.32			0.05	0.3			0.05	0.3											TLV-A4
Chloroacetyl chloride 79-04-9	0.05	0.23	0.15	0.69					0.05	0.2											
			Skin										Skin								
2-Chloroacrylonitrile 920-37-6																					MAK-3B
Chloroaniline, m-isomer 108-42-9													Skin; Sh								
Chloroaniline, o-isomer 95-51-2													Skin								
Chloroaniline, p-isomer 106-47-8													Skin; Sh								IARC-2B MAK-2

SUBSTANCE / CAS#	ACGIH® TLVs® TWA ppm	TWA mg/m³	STEL/CEIL(C) ppm	STEL/CEIL(C) mg/m³	OSHA PELs TWA ppm	TWA mg/m³	STEL/CEIL(C) ppm	STEL/CEIL(C) mg/m³	NIOSH RELs TWA ppm	TWA mg/m³	STEL/CEIL(C) ppm	STEL/CEIL(C) mg/m³	DFG MAKs TWA ppm	TWA mg/m³	PEAK/CEIL(C) ppm	PEAK/CEIL(C) mg/m³	AIHA WEELs TWA ppm	TWA mg/m³	STEL/CEIL(C) ppm	STEL/CEIL(C) mg/m³	CARCINOGENICITY CATEGORY
Chlorobenzene (Monochlorobenzene) 108-90-7	10	46		BEI	75	350							10	47	II (2)						EPA-D TLV-A3
															C						
Chlorobenzilate 510-15-6																					IARC-3
Chlorobenzotrichloride, p-isomer 5216-25-1														Skin							MAK-2
Chlorobenzylidene malononitrile, o-isomer 2698-41-1			C 0.05	C 0.39	0.05	0.4					C 0.05	C 0.4									TLV-A4
			Skin							Skin											
Chlorobromomethane (Bromochloromethane) 74-97-5	200	1060			200	1050			200	1050				Skin							EPA-D MAK-3B
1-Chlorobutane 109-69-3																					EPA-D
2-Chlorobutane 78-86-4																					EPA-D
p-Chloro-m-cresol 59-50-7														Sh							
Chlorocyclopentadiene 41851-50-7																					EPA-D

SUBSTANCE CAS#	ACGIH® TLVs TWA ppm	mg/m³	STEL/CEIL(C) ppm	mg/m³	OSHA PELs TWA ppm	mg/m³	STEL/CEIL(C) ppm	mg/m³	NIOSH RELs TWA ppm	mg/m³	STEL/CEIL(C) ppm	mg/m³	DFG MAKs TWA ppm	mg/m³	PEAK/CEIL(C) ppm	mg/m³	AIHA WEELs TWA ppm	mg/m³	STEL/CEIL(C) ppm	mg/m³	CARCINOGENICITY CATEGORY
Chlorodibromo-methane 124-48-1																					EPA-C IARC-3
3-Chloro-4-(dichloro-methyl)-5-hydroxy-2 (5H)-furanone 77439-76-0																					IARC-2B
1-Chloro-1,1-difluoro-ethane (FC-142b) 75-68-3													1000	4200	II (8) D		1000				
Chlorodifluoromethane (FC-22) 75-45-6	1000	3540							1000	3500	1250	4375	500	1800	II (8) C						IARC-3 TLV-A4
1-Chloro-2,4-dinitrobenzene 97-00-7														Sh							
Chlorodiphenyl, 42% chlorine 53469-21-9	1 Skin				1 Skin				0.001 See Pocket Guide App. A				See Chlorinated biphenyls								NIOSH-Ca
Chlorodiphenyl, 54% chlorine 11097-69-1	0.5 Skin				0.5 Skin				0.001 See Pocket Guide App. A				See Chlorinated biphenyls								NIOSH-Ca TLV-A3
bis(2-Chloroethoxy)-methane 111-91-1																					EPA-D
1-(2-Chloroethyl)-3-cy-clohexyl-1-nitrosourea (CCNU) 13010-47-4																					IARC-2A NTP-R

SUBSTANCE / CAS#	ACGIH® TLVs® TWA ppm	mg/m³	STEL/CEIL(C) ppm	mg/m³	OSHA PELs TWA ppm	mg/m³	STEL/CEIL(C) ppm	mg/m³	NIOSH RELs TWA ppm	mg/m³	STEL/CEIL(C) ppm	mg/m³	DFG MAKs TWA ppm	mg/m³	PEAK/CEIL(C) ppm	mg/m³	AIHA WEELs TWA ppm	mg/m³	STEL/CEIL(C) ppm	mg/m³	CARCINOGENICITY CATEGORY
1-(2-Chloroethyl)-3-(4-methylcyclohexyl)-1-nitrosourea (Methyl-CCNU; Semustine) 13909-09-6																					IARC-1 NTP-K
N,N-bis(2-Chloroethyl)-2-naphthylamine (Chlornaphazine) 494-03-1																					IARC-1
bis(Chloroethyl)nitro-sourea (BCNU) 154-93-8																					IARC-2A NTP-R
tris(2-Chloroethyl) phosphate 115-96-8																					IARC-3
Chlorofluoromethane (FC-31) 593-70-4																					IARC-3 MAK-2
Chloroform (Trichloromethane) 67-66-3	10	49					C 50	C 240	2* *60-min *See* Pocket Guide App. A	9.78*			0.5	2.5	II (2) Skin; C						EPA-B2; L; NL IARC-2B MAK-4 / NIOSH-Ca NTP-R TLV-A3
Chloroformic acid butyl ester 543-27-1; 592-34-7													0.2	1.1	I (2) C						
N-Chloroformyl-morpholine 15159-40-7																					MAK-2

SUBSTANCE CAS#	ACGIH® TLVs® TWA ppm	mg/m³	STEL/CEIL(C) ppm	mg/m³	OSHA PELs TWA ppm	mg/m³	STEL/CEIL(C) ppm	mg/m³	NIOSH RELs TWA ppm	mg/m³	STEL/CEIL(C) ppm	mg/m³	DFG MAKs TWA ppm	mg/m³	PEAK/CEIL(C) ppm	mg/m³	AIHA WEELs TWA ppm	mg/m³	STEL/CEIL(C) ppm	mg/m³	CARCINOGENICITY CATEGORY
bis-(2-Chloroiso-propyl)ether 39638-32-9																	3				
1,2-bis(Chloromethoxy)ethane 13483-18-6																					IARC-3
1,4-bis(Chloromethoxymethyl)benzene 56894-91-8																					IARC-3
1,2,3-tris(Chloromethoxy)propane 38571-73-2																					IARC-3
5-Chloro-2-methyl-2,3-dihydroisothiazol-3-one [26172-55-4] and 2-Methyl-2,3-dihydroiso-thiazol-3-one [2682-20-4]														0.2 I	I (2)						
													mixture in ratio 3:1 Sh; C							
bis(2-Chloro-1-methyl-ethyl)ether 108-60-1																					IARC-3
bis(Chloromethyl)ether 542-88-1	0.001	0.0047			See 29 CFR 1910.1003				See Pocket Guide App. A												EPA-A NTP-K* IARC-1* OSHA-Ca MAK-1 TLV-A1 NIOSH-Ca *includes technical grades
Chloromethyl methyl ether (CMME; Methyl chloro-methyl ether) 107-30-2				L	See 29 CFR 1910.1003				See Pocket Guide App. A												EPA-A NTP-K* IARC-1* OSHA-Ca MAK-1 TLV-A2 NIOSH-Ca *includes technical grades

SUBSTANCE CAS#	ACGIH® TLVs®				OSHA PELs				NIOSH RELs				DFG MAKs				AIHA WEELs				CARCINOGENICITY CATEGORY
	TWA		STEL/CEIL(C)		TWA		STEL/CEIL(C)		TWA		STEL/CEIL(C)		TWA		PEAK/CEIL(C)		TWA		STEL/CEIL(C)		
	ppm	mg/m³	ppm	mg/m³	ppm	mg/m³	ppm	mg/m³	ppm	mg/m³	ppm	mg/m³	ppm	mg/m³	ppm	mg/m³	ppm	mg/m³	ppm	mg/m³	
1-Chloro-2-methylpro-pene (Dimethylvinyl chloride) 513-37-1																					IARC-2B NTP-R
3-Chloro-2-methyl-propene 563-47-3																					IARC-3 MAK-3B NTP-R
1-Chloro-1-nitropropane 600-25-9	2	10			20	100			2	10											
Chloropentafluor-ethane 76-15-3	1000	6320							1000	6320											
4-Chloro-m-phenyl-enediamine 5131-60-2																					IARC-3
4-Chloro-o-phenyl-enediamine 95-83-0																					IARC-2B NTP-R
4-Chlorophenyl isocyanate 104-12-1																					MAK-3B
p-Chlorophenyl methyl sulfide 123-09-1																					EPA-D
p-Chlorophenyl methyl sulfone 98-57-7																					EPA-D

SUBSTANCE / CAS#	ACGIH® TLVs® TWA ppm	TWA mg/m³	STEL/CEIL(C) ppm	STEL/CEIL(C) mg/m³	OSHA PELs TWA ppm	TWA mg/m³	STEL/CEIL(C) ppm	STEL/CEIL(C) mg/m³	NIOSH RELs TWA ppm	TWA mg/m³	STEL/CEIL(C) ppm	STEL/CEIL(C) mg/m³	DFG MAKs TWA ppm	TWA mg/m³	PEAK/CEIL(C)	AIHA WEELs TWA ppm	TWA mg/m³	STEL/CEIL(C) ppm	STEL/CEIL(C) mg/m³	CARCINOGENICITY CATEGORY
p-Chlorophenyl methyl sulfoxide 934-73-6																				EPA-D
Chloropicrin (Trichloronitromethane) 76-06-2	0.1	0.67			0.1	0.7			0.1	0.7			0.1	0.68	I (1)					TLV-A4
β-Chloroprene (2-Chloro-1,3-butadiene) 126-99-8	10	36 Skin			25	90 Skin					C 1* *15-min See Pocket Guide App. A	C 3.6*		Skin						EPA-L NIOSH-Ca IARC-2B NTP-R MAK-2 TLV-A2
2-Chloropropane 75-29-6																50*	*OARS WEEL			
1-Chloro-2-propanol [127-00-4] and 2-Chloro-1-propanol [78-89-7]	1	4 Skin																		TLV-A4
Chloropropham 101-21-3																				IARC-3
2-Chloropropionic acid 598-78-7	0.1	0.44 Skin																		
Chloroquine 54-05-7																				IARC-3
Chlorostyrene, o-isomer 2039-87-4	50	283	75	425					50	285	75	428								

SUBSTANCE / CAS#	ACGIH® TLVs® TWA ppm	TWA mg/m³	STEL/CEIL(C) ppm	STEL/CEIL(C) mg/m³	OSHA PELs TWA ppm	TWA mg/m³	STEL/CEIL(C) ppm	STEL/CEIL(C) mg/m³	NIOSH RELs TWA ppm	TWA mg/m³	STEL/CEIL(C) ppm	STEL/CEIL(C) mg/m³	DFG MAKs TWA ppm	TWA mg/m³	PEAK/CEIL(C) ppm	PEAK/CEIL(C) mg/m³	AIHA WEELs TWA ppm	TWA mg/m³	STEL/CEIL(C) ppm	STEL/CEIL(C) mg/m³	CARCINOGENICITY CATEGORY
Chlorosulfonic acid 7790-94-5																		0.1* *OARS WEEL			
2-Chloro-1,1,1,2-tetrafluoroethane 2837-89-0																	1000				
Chlorothalonil 1897-45-6														Sh							IARC-2B MAK-3B
Chlorotoluene, o-isomer 95-49-8	50	259							50	250	75	375									
4-Chloro-o-toluidine 95-69-2														Skin; 3A							IARC-2A MAK-1 NTP-R
5-Chloro-o-toluidine 95-79-4																					IARC-3 MAK-3B
p-Chloro-o-toluidine hydrochloride 3165-93-3																					NTP-R
2-Chloro-1,1,1-trifluoroethane 75-88-7																					IARC-3
Chlorotrifluoroethylene 79-38-9																		5			

SUBSTANCE CAS#	ACGIH® TLVs® TWA ppm	mg/m³	STEL/CEIL(C) ppm	mg/m³	OSHA PELs TWA ppm	mg/m³	STEL/CEIL(C) ppm	mg/m³	NIOSH RELs TWA ppm	mg/m³	STEL/CEIL(C) ppm	mg/m³	DFG MAKs TWA ppm	mg/m³	PEAK/CEIL(C)	AIHA WEELs TWA ppm	mg/m³	STEL/CEIL(C) ppm	mg/m³	CARCINOGENICITY CATEGORY
Chlorotrifluoromethane (FC-13) 75-72-9													1000	4300	II (8) D					
trans-1-Chloro-3,3,3-trifluoropropylene 102687-65-0																800*		*OARS WEEL		
Chlorozotocin 54749-90-5																				IARC-2A NTP-R
Chlorpromazine 50-53-3															SP					
Chlorpyrifos 2921-88-2	0.1 IFV Skin; BEI_A								0.2 Skin		0.6									TLV-A4
Cholesterol 57-88-5																				IARC-3
Chromic acid [7738-94-5] and chromates	*See* Chromium(VI) inorganic compounds, water-soluble						C 0.1*	*as CrO₃	0.0002* *as Cr *See* Pocket Guide Apps. A and C				*See* Chromium(VI) inorganic compounds, water-soluble							NIOSH-Ca
Chromite ore processing (Chromate), as Cr	(0.05) NIC-withdraw line entry from *TLVs® and BEIs®* book; *see* Chromium and inorganic compounds (III, VI)																			(TLV-A1)
Chromium(II) inorganic compounds, as Cr					0.5				0.5 *See* Pocket Guide App. C											

SUBSTANCE / CAS#	ACGIH® TLVs® TWA ppm	TWA mg/m³	STEL/CEIL(C) ppm	STEL/CEIL(C) mg/m³	OSHA PELs TWA ppm	TWA mg/m³	STEL/CEIL(C) ppm	STEL/CEIL(C) mg/m³	NIOSH RELs TWA ppm	TWA mg/m³	STEL/CEIL(C) ppm	STEL/CEIL(C) mg/m³	DFG MAKs TWA ppm	TWA mg/m³	PEAK/CEIL(C) ppm	PEAK/CEIL(C) mg/m³	AIHA WEELs TWA ppm	TWA mg/m³	STEL/CEIL(C) ppm	STEL/CEIL(C) mg/m³	CARCINOGENICITY CATEGORY
Chromium(III) [16065-83-1] inorgainc compounds, as Cr 7440-47-3		(0.5) NIC-0.003 **I*** *as Cr(III) NIC-DSEN**; RSEN** **water-soluble compounds only				0.5				0.5 *See* Pocket Guide App. C					Sh* *does not apply to Cr(III) oxide and similar poorly soluble Cr(III) cmpds							EPA-D; CBD IARC-3* TLV-A4 *organic and inorganic compounds
Chromium(VI) [18540-29-9] inorganic compounds, water-soluble		(0.05*) NIC-0.0002 **I*** *as Cr(VI) NIC-Skin; DSEN, RSEN BEI	NIC-0.0005 **I***		0.005* *as Cr(VI) *See* 29 CFR 1910.1026				0.0002* *as Cr(VI) *See* Pocket Guide Apps. A and C				Inhalable fraction Skin*; Sh**; 2 *the chromates of Ba, Pb, Sr, and Zn are not designated with Skin **BaCrO₄ and PbCrO₄ are not designated with Sh								EPA-A*; K*;　NIOSH-Ca D**; CBD**　NTP-K IARC-1　TLV-A1 MAK-1 *inhalation **oral
Chromium(VI) [18540-29-9] inorganic compounds, insoluble		(0.01*) NIC-0.0002 **I*** *as Cr(VI) NIC-DSEN; RSEN	NIC-0.0005 **I***		0.005* *as Cr(VI) *See* 29 CFR 1910.1026				0.0002* *as Cr(VI) *See* Pocket Guide Apps. A and C				Inhalable fraction Skin*; Sh**; 2 *the chromates of Ba, Pb, Sr, and Zn are not designated with Skin **BaCrO₄ and PbCrO₄ are not designated with Sh								EPA-A*; K*;　NIOSH-Ca D**; CBD**　NTP-K IARC-1　TLV-A1 MAK-1 *inhalation **oral
Chromium metal 7440-47-3		(0.5) 0.5 **I*** *as Cr(0)			1				0.5 *See* Pocket Guide App. C											EPA-A*; K*;　IARC-3 D**; CBD**　TLV-A4 *inhalation **oral
Chromyl chloride 14977-61-8	(0.025)	(0.16)	NIC-0.00025* NIC-0.0005* NIC-0.001* NIC-0.0002* *as Cr(VI); **IFV** NIC-A1 NIC-Skin; DSEN; RSEN						0.001* *as Cr *See* Pocket Guide Apps. A and C				*See* Chromium(VI) inorganic compounds, water-soluble							IARC-1 MAK-1 NIOSH-Ca NTP-K
Chrysene 218-01-9		L; BEI_P			0.2 *See* Coal tar pitch volatiles				0.1* *Cyclohexane-extractable fraction *See* Pocket Guide Apps. A and C				Skin								EPA-B2　NIOSH-Ca IARC-2B　TLV-A3 MAK-2
Chrysoidine 532-82-1																					IARC-3

SUBSTANCE CAS#	ACGIH® TLVs® TWA ppm	mg/m³	STEL/CEIL(C) ppm	mg/m³	OSHA PELs TWA ppm	mg/m³	STEL/CEIL(C) ppm	mg/m³	NIOSH RELs TWA ppm	mg/m³	STEL/CEIL(C) ppm	mg/m³	DFG MAKs TWA ppm	mg/m³	PEAK/CEIL(C) ppm	mg/m³	AIHA WEELs TWA ppm	mg/m³	STEL/CEIL(C) ppm	mg/m³	CARCINOGENICITY CATEGORY
CI Acid Orange 3 6373-74-6																					IARC-3
CI Acid Red 114 6459-94-5																					IARC-2B
CI Basic Red 9 569-61-9																					IARC-2B NTP-R
CI Direct Blue 15 2429-74-5																					IARC-2B
Cimetidine 51481-61-9																					IARC-3
Cinnamaldehyde 104-55-2															Sh							
Cinnamyl alcohol 104-54-1															Sh							
Cinnamyl anthranilate 87-29-6																					IARC-3
CI Pigment Red 3 2425-85-6																					IARC-3

SUBSTANCE / CAS#	ACGIH® TLVs® TWA ppm	TWA mg/m³	STEL/CEIL(C) ppm	STEL/CEIL(C) mg/m³	OSHA PELs TWA ppm	TWA mg/m³	STEL/CEIL(C) ppm	STEL/CEIL(C) mg/m³	NIOSH RELs TWA ppm	TWA mg/m³	STEL/CEIL(C) ppm	STEL/CEIL(C) mg/m³	DFG MAKs TWA ppm	TWA mg/m³	PEAK/CEIL(C) ppm	PEAK/CEIL(C) mg/m³	AIHA WEELs TWA ppm	TWA mg/m³	STEL/CEIL(C) ppm	STEL/CEIL(C) mg/m³	CARCINOGENICITY CATEGORY
Cisplatin 15663-27-1																					IARC-2A NTP-R
Citral 5392-40-5	5 **IFV**		32 **IFV**																		TLV-A4
Skin; DSEN																					
Citrinin 518-75-2																					IARC-3
Citrus Red No. 2 6358-53-8																					IARC-2B
Clofibrate 637-07-0																					IARC-3
Clomiphene citrate 50-41-9																					IARC-3
Clopidol 2971-90-6	3 **IFV**				15*; 5**				10*; 5**		20*										TLV-A4
					*Total dust **Respirable fraction				*Total dust **Respirable fraction												
Coal dust, anthracite; bituminous or lignite	0.4 **R*** 0.9 **R****				2.4*				1* 0.9**												IARC-3 MAK-3B* TLV-A4

Coal dust OSHA: $\dfrac{10\ \text{mg/m}^{3**}}{\%\ SiO_2 + 2}$

NIOSH Coal dust notes: * measured according to MSHA method (CPSU) ** measured according to SO/CEN/ACGIH® criteria *See* Pocket Guide App. C

ACGIH notes: *Anthracite **Bituminous or lignite

OSHA notes: *< 5% SiO_2, resp. Quartz fraction ** ≥ 5% SiO_2, resp. Quartz fraction

Carcinogenicity note: * Coal mine dust, as respirable fraction

SUBSTANCE / CAS#	ACGIH® TLVs® TWA ppm	TWA mg/m³	STEL/CEIL(C) ppm	STEL/CEIL(C) mg/m³	OSHA PELs TWA ppm	TWA mg/m³	STEL/CEIL(C) ppm	STEL/CEIL(C) mg/m³	NIOSH RELs TWA ppm	TWA mg/m³	STEL/CEIL(C) ppm	STEL/CEIL(C) mg/m³	DFG MAKs TWA ppm	TWA mg/m³	PEAK/CEIL(C) ppm	PEAK/CEIL(C) mg/m³	AIHA WEELs TWA ppm	TWA mg/m³	STEL/CEIL(C) ppm	STEL/CEIL(C) mg/m³	CARCINOGENICITY CATEGORY
Coal gasification																					IARC-1
Coal-tar distillation																					IARC-1
Coal-tar pitch																					IARC-1 NTP-K
Coal tar pitch volatiles, as benzene soluble aerosol 65996-93-2	0.2 BEI_p				0.2				0.1* *Cyclohexane-extractable fraction *See* Pocket Guide Apps. A and C												IARC-1 NIOSH-Ca NTP-K TLV-A1
Cobalt, alloys													Cobalt alloys containing bio-available Cobalt, *see* Cobalt and compounds								
Cobalt [7440-48-4] and compounds													as Inhalable fraction Skin; Sah; 3A								IARC-2B MAK-2 NTP-R* *that releases Cobalt ions *in vivo*
Cobalt [7440-48-4] and inorganic compounds, as Co	(0.02) NIC-0.02 **I** NIC-Skin; DSEN; RSEN (BEI)				0.1* *for metal dust and fume				0.05* *for metal dust and fume												TLV-A3
Cobalt [7440-48-4] with tungsten carbide [12070-12-1]	BEI																				IARC-2A NTP-R* *powders and hard metals
Cobalt carbonyl, as Co 10210-68-1	0.1								0.1												IARC-2B

SUBSTANCE / CAS#	ACGIH® TLVs® TWA ppm	mg/m³	STEL/CEIL(C) ppm	mg/m³	OSHA PELs TWA ppm	mg/m³	STEL/CEIL(C) ppm	mg/m³	NIOSH RELs TWA ppm	mg/m³	STEL/CEIL(C) ppm	mg/m³	DFG MAKs TWA ppm	mg/m³	PEAK/CEIL(C) ppm	mg/m³	AIHA WEELs TWA ppm	mg/m³	STEL/CEIL(C) ppm	mg/m³	CARCINOGENICITY CATEGORY
Cobalt hydrocarbonyl, as Co 16842-03-8	0.1								0.1												IARC-2B
Cobalt sulfate 10124-43-3													*See* Cobalt and compounds								IARC-2B* NTP-R *and other soluble Cobalt (II) salts
Cocamide diethanola-mine 68603-42-9																					IARC-2B
Coke oven emissions					0.15* *Benzene-soluble fraction *See* 29 CFR 1910.1029				0.2* *Benzene-soluble fraction *See* Pocket Guide Apps. A and C												EPA-A NTP-K MAK-1 OSHA-Ca NIOSH-Ca
Coke production																					IARC-1
Copper, dusts and mists, as Cu 7440-50-8	1				1				1												EPA-D
Copper, fume, as Cu 7440-50-8	0.2				0.1				0.1												EPA-D
Copper [7440-50-8] and its inorganic compounds													0.01 **R**		II (2) C						EPA-D
Copper-8-hydroxy-quinoline 10380-28-6																					IARC-3

SUBSTANCE CAS#	ACGIH® TLVs® TWA ppm	mg/m³	STEL/CEIL(C) ppm	mg/m³	OSHA PELs TWA ppm	mg/m³	STEL/CEIL(C) ppm	mg/m³	NIOSH RELs TWA ppm	mg/m³	STEL/CEIL(C) ppm	mg/m³	DFG MAKs TWA ppm	mg/m³	PEAK/CEIL(C) ppm	mg/m³	AIHA WEELs TWA ppm	mg/m³	STEL/CEIL(C) ppm	mg/m³	CARCINOGENICITY CATEGORY
Coronene 191-07-1																					IARC-3
Cotton dust, in textile mill waste house operations, in yarn manufacturing, or from "lower-grade washed cotton"					0.5*																
					*Lint-free respirable dust, as measured by vert. elutriator. See 29 CFR 1910.1043																
Cotton dust, in textile slashing and weaving operations					0.75*																
					*Lint-free respirable dust, as measured by vert. elutriator. See 29 CFR 1910.1043																
Cotton dust, in yarn manufacturing and cotton washing operations					0.2*																
					*Lint-free respirable dust, as measured by vert. elutriator. See 29 CFR 1910.1043																
Cotton dust, raw, untreated		0.1 T			1*				< 0.2					1.5 I		I (1)					TLV-A4
					*Resp. dust, measured by vert. elutriator (cotton waste processing oper.)				See Pocket Guide App. C							C					
Coumaphos 56-72-4		0.05 IFV																			TLV-A4
		Skin; BEI_A																			
Coumarin 91-64-5																					IARC-3
Creosotes 8001-58-9																					EPA-B1 IARC-2A

SUBSTANCE / CAS#	ACGIH® TLVs® TWA ppm	mg/m³	STEL/CEIL(C) ppm	mg/m³	OSHA PELs TWA ppm	mg/m³	STEL/CEIL(C) ppm	mg/m³	NIOSH RELs TWA ppm	mg/m³	STEL/CEIL(C) ppm	mg/m³	DFG MAKs TWA ppm	mg/m³	PEAK/CEIL(C) ppm	mg/m³	AIHA WEELs TWA ppm	mg/m³	STEL/CEIL(C) ppm	mg/m³	CARCINOGENICITY CATEGORY
Cresidine, m-isomer 102-50-1																					IARC-3
Cresidine, p-isomer (5-Methyl-o-anisidine) 120-71-8																					IARC-2B MAK-2 NTP-R
Cresol, all isomers 95-48-7; 106-44-5; 108-39-4; 1319-77-3	20 IFV Skin				5	22 Skin			2.3	10			Skin								EPA-C* MAK-3A TLV-A4 *o, m, p isomers only
Cresyl glycidyl ether, o-isomer [2210-79-9], or mixture [26447-14-3]													Sh								
Crotonaldehyde 123-73-9; 4170-30-3			C 0.3 Skin	C 0.86	2	6			2	6 *See* Pocket Guide App. C			Skin; 3B								EPA-C IARC-3 MAK-3B TLV-A3
Crufomate 299-86-5	5 BEI_A								5		20										TLV-A4
Cumene 98-82-8	(50) NIC-0.1 NIC-A2	(246) NIC-0.5			50	245 Skin			50	245 Skin			10 Skin; C	50	II (4)						EPA-CBD; D IARC-2B MAK-3B NTP-R
Cumene hydroperoxide 80-15-9															Skin		1	6			
Cupferron 135-20-6																					NTP-R

SUBSTANCE / CAS#	ACGIH® TLVs® TWA ppm	TWA mg/m³	STEL/CEIL(C) ppm	STEL/CEIL(C) mg/m³	OSHA PELs TWA ppm	TWA mg/m³	STEL/CEIL(C) ppm	STEL/CEIL(C) mg/m³	NIOSH RELs TWA ppm	TWA mg/m³	STEL/CEIL(C) ppm	STEL/CEIL(C) mg/m³	DFG MAKs TWA ppm	TWA mg/m³	PEAK/CEIL(C)	AIHA WEELs TWA ppm	TWA mg/m³	STEL/CEIL(C) ppm	STEL/CEIL(C) mg/m³	CARCINOGENICITY CATEGORY
Cyanamide 420-04-2	2								2				0.2	0.35*	II (1)					
													*can also occur as vapor and aerosol Skin; Sh; C							
Cyanazine 21725-46-2	NIC-0.1 I NIC-A3																			
Cyanides, as CN					5								2 I		II (1)					
					Skin								Skin; C							
Cyanide, free 57-12-5																				EPA-D
Cyanoacrylates 137-05-3; 7085-85-0	NIC-0.2 NIC-1		NIC-1 NIC-5.1																	
	NIC-DSEN; RSEN																			
Cyanogen 460-19-5			C 5	C 10.6					10	20			5	11	II (2)					
													Skin; D							
Cyanogen bromide 506-68-3			C 0.3	C 1.3																
Cyanogen chloride 506-77-4			C 0.3	C 0.75							C 0.3	C 0.6								
Cycasin 14901-08-7																				IARC-2B

SUBSTANCE / CAS#	ACGIH® TLVs® TWA ppm	mg/m³	STEL/CEIL(C) ppm	mg/m³	OSHA PELs TWA ppm	mg/m³	STEL/CEIL(C) ppm	mg/m³	NIOSH RELs TWA ppm	mg/m³	STEL/CEIL(C) ppm	mg/m³	DFG MAKs TWA ppm	mg/m³	PEAK/CEIL(C) ppm	mg/m³	AIHA WEELs TWA ppm	mg/m³	STEL/CEIL(C) ppm	mg/m³	CARCINOGENICITY CATEGORY
Cyclochlorotine 12663-46-6																					IARC-3
Cyclohexane 110-82-7	100	344			300	1050			300	1050			200	700	II (4) D						EPA-I
Cyclohexanol 108-93-0	50	206 Skin			50	200			50	200 Skin				Skin							
Cyclohexanone 108-94-1	20	Skin	50		50	200			25	100 Skin				Skin							IARC-3 MAK-3B TLV-A3
Cyclohexene 110-83-8	300	1010			300	1015			300	1015											
Cyclohexylamine 108-91-8	10	41							10	40			2	8.2 a momentary value of 21 mg/m³ should not be exceeded C	I (2)						TLV-A4
N-Cyclohexyl-2-benz-othiazolesulfenamide 95-33-0														Sh							
N-Cyclohexylhydroxy-diazene-1-oxide, copper salt 15627-09-5													0.05 R	Skin; C	II (2)						
Cyclohexylhydroxydia-zene-1-oxide, potas-sium salt 66603-10-9													10 I	Skin; D	II (2)						

SUBSTANCE CAS#	ACGIH® TLVs® TWA ppm	ACGIH® TLVs® TWA mg/m³	ACGIH® TLVs® STEL/CEIL(C) ppm	ACGIH® TLVs® STEL/CEIL(C) mg/m³	OSHA PELs TWA ppm	OSHA PELs TWA mg/m³	OSHA PELs STEL/CEIL(C) ppm	OSHA PELs STEL/CEIL(C) mg/m³	NIOSH RELs TWA ppm	NIOSH RELs TWA mg/m³	NIOSH RELs STEL/CEIL(C) ppm	NIOSH RELs STEL/CEIL(C) mg/m³	DFG MAKs TWA ppm	DFG MAKs TWA mg/m³	DFG MAKs PEAK/CEIL(C) ppm	DFG MAKs PEAK/CEIL(C) mg/m³	AIHA WEELs TWA ppm	AIHA WEELs TWA mg/m³	AIHA WEELs STEL/CEIL(C) ppm	AIHA WEELs STEL/CEIL(C) mg/m³	CARCINOGENICITY CATEGORY
Cyclohexylmercaptan 1569-69-3											C 0.5* *15-min	C 2.4*									
N-Cyclohexyl-N′-phenyl-p-phenylenediamine 101-87-1															Sh						
Cyclonite (RDX) 121-82-4		0.5 Skin								1.5 Skin	3										EPA-C TLV-A4
4H-Cyclopenta[def]-chrysene 202-98-2																					IARC-3
Cyclopentadiene 542-92-7	75	203			75	200			75	200											
Cyclopentane 287-92-3	600	1720							600	1720											
Cyclopenta[cd]pyrene 27208-37-3															Skin; 3B						IARC-2A MAK-2
5,6-Cyclopenteno-1,2-benzanthracene 7099-43-6																					IARC-3
Cyclophosphamide 50-18-0; 6055-19-2																					IARC-1 NTP-K* *CAS: 50-18-0

SUBSTANCE / CAS#	ACGIH® TLVs TWA ppm	mg/m³	STEL/CEIL(C) ppm	mg/m³	OSHA PELs TWA ppm	mg/m³	STEL/CEIL(C) ppm	mg/m³	NIOSH RELs TWA ppm	mg/m³	STEL/CEIL(C) ppm	mg/m³	DFG MAKs TWA ppm	mg/m³	PEAK/CEIL(C) ppm	mg/m³	AIHA WEELs TWA ppm	mg/m³	STEL/CEIL(C) ppm	mg/m³	CARCINOGENICITY CATEGORY
Cyclosporine 79217-60-0																					IARC-1
Cyclosporin A 59865-13-3																					IARC-1 NTP-K
Cyfluthrin 68359-37-5														0.01 I	I (1) C						
Cyhexatin (Tricyclohexyltin hydroxide) 13121-70-5	5				0.1* *as Sn				5												TLV-A4
2,4-D (2,4-Dichloro-phenoxyacetic acid) 94-75-7	10 I				10				10				2 I including salts and esters Skin; C	II (2)							TLV-A4
D and C Red No. 9 (Benzenesulfonic acid; 5-Chloro-2-[(2-hydroxy-1-naphthalenyl)azo]-4-methylbenzenesulfonic acid, barium salt (2:1)) 5160-02-1																		1 as Barium salt (2:1)			IARC-3
Dacarbazine 4342-03-4																					IARC-2B NTP-R
Dantron (Chrysazin; 1,8-Dihydroxyanthra-quinone) 117-10-2																					IARC-2B NTP-R

SUBSTANCE CAS#	ACGIH® TLVs® TWA ppm	mg/m³	STEL/CEIL(C) ppm	mg/m³	OSHA PELs TWA ppm	mg/m³	STEL/CEIL(C) ppm	mg/m³	NIOSH RELs TWA ppm	mg/m³	STEL/CEIL(C) ppm	mg/m³	DFG MAKs TWA ppm	mg/m³	PEAK/CEIL(C)	AIHA WEELs TWA ppm	mg/m³	STEL/CEIL(C) ppm	mg/m³	CARCINOGENICITY CATEGORY
Dapsone 80-08-0																				IARC-3
Daunomycin 20830-81-3																				IARC-2B
Dawsonite, fibrous dust 12011-76-6																				MAK-2
DDT (Dichlorodiphenyl-trichloroethane) 50-29-3		1				1				0.5				1 I	II (8)					EPA-B2 TLV-A3 IARC-2A NIOSH-Ca NTP-R
						Skin				*See* Pocket Guide App. A				Skin						
Decaborane 17702-41-9	0.05	0.25	0.15	0.75	0.05	0.3			0.05	0.3	0.15	0.9	0.05	0.25	II (2)					
		Skin				Skin				Skin				Skin						
Decabromodiphenyl oxide 1163-19-5																	5			EPA-S IARC-3
Decahydronaphtha-lene 91-17-8													5	29*	II (2)					
													*can also occur as vapor and aerosol D						
Decamethylcyclo-pentasiloxane 541-02-6																10*				
																*OARS WEEL				
1-Decene 872-05-9																100				

SUBSTANCE CAS#	ACGIH® TLVs® TWA ppm	ACGIH® TLVs® TWA mg/m³	ACGIH® TLVs® STEL/CEIL(C) ppm	ACGIH® TLVs® STEL/CEIL(C) mg/m³	OSHA PELs TWA ppm	OSHA PELs TWA mg/m³	OSHA PELs STEL/CEIL(C) ppm	OSHA PELs STEL/CEIL(C) mg/m³	NIOSH RELs TWA ppm	NIOSH RELs TWA mg/m³	NIOSH RELs STEL/CEIL(C) ppm	NIOSH RELs STEL/CEIL(C) mg/m³	DFG MAKs TWA ppm	DFG MAKs TWA mg/m³	DFG MAKs PEAK/CEIL(C) ppm	DFG MAKs PEAK/CEIL(C) mg/m³	AIHA WEELs TWA ppm	AIHA WEELs TWA mg/m³	AIHA WEELs STEL/CEIL(C) ppm	AIHA WEELs STEL/CEIL(C) mg/m³	CARCINOGENICITY CATEGORY
Decyl alcohol 112-30-1													10	66* *can also occur as vapor and aerosol C	I (1)						
Decylmercaptan 143-10-2											C 0.5* *15-min	C 3.6*									
2-Dehydrolinalool 29171-20-8																	2				
Deltamethrin 52918-63-5																					IARC-3
Demeton 8065-48-3		0.05 IFV Skin; BEI$_A$				0.1 Skin				0.1 Skin				Skin							
Demeton-S-methyl 919-86-8		0.05 IFV Skin; DSEN; BEI$_A$																			TLV-A4
Diacetone alcohol (4-Hydroxy-4-methyl-2-pentanone) 123-42-2	50	238			50	240			50	240			20	96 Skin; D	I (2)						
Diacetyl 431-03-8	0.01	0.04	0.02	0.07									0.02	0.071 Skin; Sh; C	II (1)						MAK-3B TLV-A4
Diacetylaminoazo-toluene 83-63-6																					IARC-3

SUBSTANCE / CAS#	ACGIH® TLVs® TWA ppm	mg/m³	STEL/CEIL(C) ppm	mg/m³	OSHA PELs TWA ppm	mg/m³	STEL/CEIL(C) ppm	mg/m³	NIOSH RELs TWA ppm	mg/m³	STEL/CEIL(C) ppm	mg/m³	DFG MAKs TWA ppm	mg/m³	PEAK/CEIL(C)	AIHA WEELs TWA ppm	mg/m³	STEL/CEIL(C) ppm	mg/m³	CARCINOGENICITY CATEGORY
N,N′-Diacetylbenzidine 613-35-4																				IARC-2B
Diallate 2303-16-4																				IARC-3
Diallylamine 124-02-7																1				
														Skin						
2,4-Diaminoanisole 615-05-4										and its salts; minimize occupational exposure (especially skin exposures) *See* Pocket Guide App. A					Skin					IARC-2B MAK-2 NIOSH-Ca
2,4-Diaminoanisole sulfate 39156-41-7																				NTP-R
3,3′-Diaminobenzidine [91-95-2] **and 3,3′-Diaminobenzidine tetrahydrochloride** [7411-49-6]																				MAK-3B
1,2-Diamino-4-nitrobenzene 99-56-9																				IARC-3
2,4-Diaminotoluene (Toluene-2,4-diamine) 95-80-7										all isomers *See* Pocket Guide App. A					Skin; Sh	0.005	and mixed isomers CAS: 25376-45-8 Skin			IARC-2B NTP-R MAK-2 NIOSH-Ca

SUBSTANCE / CAS#	ACGIH® TLVs® TWA ppm	mg/m³	STEL/CEIL(C) ppm	mg/m³	OSHA PELs TWA ppm	mg/m³	STEL/CEIL(C) ppm	mg/m³	NIOSH RELs TWA ppm	mg/m³	STEL/CEIL(C) ppm	mg/m³	DFG MAKs TWA ppm	mg/m³	PEAK/CEIL(C) ppm	mg/m³	AIHA WEELs TWA ppm	mg/m³	STEL/CEIL(C) ppm	mg/m³	CARCINOGENICITY CATEGORY
2,5-Diaminotoluene (Toluene-2,5-diamine) 95-70-5															Sh						IARC-3
Dianisidine-based dyes									minimize exposure; handle with caution *See* Pocket Guide App. C												NIOSH-Ca
Diazepam 439-14-5																					IARC-3
Diazinon 333-41-5	0.01 **IFV** Skin; BEI_A								0.1 Skin				0.1 **I** Skin; C		II (2)						IARC-2A TLV-A4
Diazoaminobenzene 136-35-6																					NTP-R
Diazomethane 334-88-3	0.2	0.34			0.2	0.4			0.2	0.4											IARC-3 MAK-2 TLV-A2
Dibenz[a,h]acridine 226-36-8																					IARC-2B NTP-R
Dibenz[a,j]acridine 224-42-0																					IARC-2A NTP-R
Dibenz[c,h]acridine 224-53-3																					IARC-2B

SUBSTANCE CAS#	ACGIH® TLVs® TWA ppm	mg/m³	STEL/CEIL(C) ppm	mg/m³	OSHA PELs TWA ppm	mg/m³	STEL/CEIL(C) ppm	mg/m³	NIOSH RELs TWA ppm	mg/m³	STEL/CEIL(C) ppm	mg/m³	DFG MAKs TWA ppm	mg/m³	PEAK/CEIL(C)	AIHA WEELs TWA ppm	mg/m³	STEL/CEIL(C) ppm	mg/m³	CARCINOGENICITY CATEGORY
Dibenz[a,c]anthracene 215-58-7																				IARC-3
Dibenz[a,h]anthracene 53-70-3															Skin; 3A					EPA-B2 IARC-2A MAK-2 NTP-R
Dibenz[a,j]anthracene 224-41-9																				IARC-3
7H-Dibenzo[c,g]-carbazole 194-59-2																				IARC-2B NTP-R
Dibenzo-p-dioxin 262-12-4																				IARC-3
Dibenzo[a,e]fluoran-thene 5385-75-1																				IARC-3
13H-Dibenzo[a,g]-fluorene 207-83-0																				IARC-3
Dibenzofuran 132-64-9																				EPA-D
Dibenzo[h,rst]penta-phene 192-47-2																				IARC-3

SUBSTANCE / CAS#	ACGIH® TLVs® TWA ppm	mg/m³	STEL/CEIL(C) ppm	mg/m³	OSHA PELs TWA ppm	mg/m³	STEL/CEIL(C) ppm	mg/m³	NIOSH RELs TWA ppm	mg/m³	STEL/CEIL(C) ppm	mg/m³	DFG MAKs TWA ppm	mg/m³	PEAK/CEIL(C) ppm	mg/m³	AIHA WEELs TWA ppm	mg/m³	STEL/CEIL(C) ppm	mg/m³	CARCINOGENICITY CATEGORY
Dibenzo[a,e]pyrene 192-65-4													Skin; 3B								IARC-3 MAK-2 NTP-R
Dibenzo[a,h]pyrene 189-64-0													Skin; 3B								IARC-2B MAK-2 NTP-R
Dibenzo[a,i]pyrene 189-55-9													Skin; 3B								IARC-2B MAK-2 NTP-R
Dibenzo[a,l]pyrene 191-30-0													Skin; 3B								IARC-2A MAK-2 NTP-R
Dibenzo[e,l]pyrene 192-51-8																					IARC-3
2,2'-Dibenzothiazyl disulfide 120-78-5													Sh								
Dibenzothiophene 132-65-0																					IARC-3
Diborane 19287-45-7	0.1	0.11			0.1	0.1			0.1	0.1											
Dibromoacetic acid 631-64-1																					IARC-2B

SUBSTANCE CAS#	ACGIH® TLVs® TWA ppm	ACGIH® TLVs® TWA mg/m³	ACGIH® TLVs® STEL/CEIL(C) ppm	ACGIH® TLVs® STEL/CEIL(C) mg/m³	OSHA PELs TWA ppm	OSHA PELs TWA mg/m³	OSHA PELs STEL/CEIL(C) ppm	OSHA PELs STEL/CEIL(C) mg/m³	NIOSH RELs TWA ppm	NIOSH RELs TWA mg/m³	NIOSH RELs STEL/CEIL(C) ppm	NIOSH RELs STEL/CEIL(C) mg/m³	DFG MAKs TWA ppm	DFG MAKs TWA mg/m³	DFG MAKs PEAK/CEIL(C) ppm	DFG MAKs PEAK/CEIL(C) mg/m³	AIHA WEELs TWA ppm	AIHA WEELs TWA mg/m³	AIHA WEELs STEL/CEIL(C) ppm	AIHA WEELs STEL/CEIL(C) mg/m³	CARCINOGENICITY CATEGORY
Dibromoacetonitrile 3252-43-5																					IARC-2B
1,2-Dibromo-3-chloro-propane (DBCP) 96-12-8					0.001	*See* 29 CFR 1910.1044				*See* Pocket Guide App. A				Skin; 2							IARC-2B OSHA-Ca MAK-2 NIOSH-Ca NTP-R
2,2-Dibromo-2-cyan-acetamide 10222-01-2														Sh							
Dibromodichloro-methane 594-18-3																					EPA-D
1,2-Dibromo-2,4-dicyanobutane 35691-65-7														Sh							
p,p′-Dibromodiphenyl ether 2050-47-7																					EPA-D
2,3-Dibromo-1-propanol 96-13-9																					IARC-2B NTP-R
tris(2,3-Dibromopropyl)-phosphate 126-72-7																					IARC-2A NTP-R
Dibutylamine 111-92-2																		C 5 Skin			

SUBSTANCE / CAS#	ACGIH® TLVs® TWA ppm	TWA mg/m³	STEL/CEIL(C) ppm	STEL/CEIL(C) mg/m³	OSHA PELs TWA ppm	TWA mg/m³	STEL/CEIL(C) ppm	STEL/CEIL(C) mg/m³	NIOSH RELs TWA ppm	TWA mg/m³	STEL/CEIL(C) ppm	STEL/CEIL(C) mg/m³	DFG MAKs TWA ppm	TWA mg/m³	PEAK/CEIL(C) ppm	PEAK/CEIL(C) mg/m³	AIHA WEELs TWA ppm	TWA mg/m³	STEL/CEIL(C) ppm	STEL/CEIL(C) mg/m³	CARCINOGENICITY CATEGORY
2-N-Dibutylaminoethanol 102-81-8	0.5	3.5							2	14											
	Skin; BEI_A								Skin												
3,5-Di-tert-butyl-4-hydroxyphenyl pro-pionic acid octadecyl ester 2082-79-3														20 I	II (2)						
														C							
Dibutyl phenyl phosphate 2528-36-1	0.3	3.5																			
	Skin; BEI_A																				
Dibutyl phosphate 107-66-4	0.6 IFV	5 IFV			1	5			1	5	2	10									MAK-3A* *and its technical mixture
	Skin																				
Dibutyl phthalate 84-74-2		5				5				5			0.05	0.58*	I (2)						EPA-D MAK-3B
													*can also occur as vapor and aerosol							
													C								
Di-n-butyltin compounds, as Sn													0.004*	0.02 I	I (1)						MAK-4
													Skin**; B							
													*can also be found as vapor **for n-butyltin cmpds whose organic ligands were already designated "Sa" or "Sh", these designations also apply							
Dicarboxylic acid [C₄–C₆] dimethyl ester, mixture 95481-62-2													0.75	5	I (1)						
													C								

SUBSTANCE CAS#	ACGIH® TLVs® TWA ppm	mg/m³	STEL/CEIL(C) ppm	mg/m³	OSHA PELs TWA ppm	mg/m³	STEL/CEIL(C) ppm	mg/m³	NIOSH RELs TWA ppm	mg/m³	STEL/CEIL(C) ppm	mg/m³	DFG MAKs TWA ppm	mg/m³	PEAK/CEIL(C) ppm	mg/m³	AIHA WEELs TWA ppm	mg/m³	STEL/CEIL(C) ppm	mg/m³	CARCINOGENICITY CATEGORY
Dichloroacetic acid 79-43-6	0.5	2.64 Skin																			EPA-L IARC-2B MAK-3A TLV-A3
Dichloroacetonitrile 3018-12-0																					IARC-3
Dichloroacetylene 7572-29-4			C 0.1	C 0.39							C 0.1 *See* Pocket Guide App. A	C 0.4									IARC-3 MAK-2 NIOSH-Ca TLV-A3
3,4-Dichloroaniline 95-76-1														Skin; Sh							
Dichlorobenzene, m-isomer 541-73-1													2	12	II (2) C						EPA-D IARC-3
Dichlorobenzene, o-isomer (1,2-Dichloro-benzene) 95-50-1	25	150	50	301			C 50	C 300			C 50	C 300	10	61 Skin; C	II (2)						EPA-D IARC-3 TLV-A4
Dichlorobenzene, p-isomer (1,4-Dichloro-benzene) 106-46-7	10	60			75	450			*See* Pocket Guide App. A					Skin; 3B							IARC-2B NTP-R MAK-2 TLV-A3 NIOSH-Ca
3,3'-Dichlorobenzidine 91-94-1		Skin; L			*See* 29 CFR 1910.1003				and its salts *See* Pocket Guide App. A					Skin							EPA-B2 NTP-R IARC-2B OSHA-Ca MAK-2 TLV-A3 NIOSH-Ca
3,3'-Dichlorobenzidine dihydrochloride 612-83-9																					NTP-R

SUBSTANCE / CAS#	ACGIH® TLVs TWA ppm	TWA mg/m³	STEL/CEIL(C) ppm	STEL/CEIL(C) mg/m³	OSHA PELs TWA ppm	TWA mg/m³	STEL/CEIL(C) ppm	STEL/CEIL(C) mg/m³	NIOSH RELs TWA ppm	TWA mg/m³	STEL/CEIL(C) ppm	STEL/CEIL(C) mg/m³	DFG MAKs TWA ppm	TWA mg/m³	PEAK/CEIL(C) ppm	PEAK/CEIL(C) mg/m³	AIHA WEELs TWA ppm	TWA mg/m³	STEL/CEIL(C) ppm	STEL/CEIL(C) mg/m³	CARCINOGENICITY CATEGORY
1,4-Dichloro-2-butene 764-41-0	0.005	0.025 Skin												Skin; 3A							MAK-2 TLV-A2
1,4-Dichlorobutene, trans-isomer 110-57-6																					IARC-3
3,3′-Dichloro-4,4′-di-aminodiphenyl ether 28434-86-8																					IARC-2B
Dichlorodifluoromethane (FC-12) 75-71-8	1000	4950			1000	4950			1000	4950			1000	5000	II (2) C						TLV-A4
1,3-Dichloro-5,5-di-methylhydantoin 118-52-5		0.2		0.4		0.2				0.2		0.4									
p,p′-Dichlorodiphenyl-dichloroethane (DDD) 72-54-8																					EPA-B2
p,p′-Dichlorodiphenyl-dichloroethylene (DDE) 72-55-9																					EPA-B2
1,1-Dichloroethane (Ethylidene chloride) 75-34-3	100	405			100	400			100	400 *See* Pocket Guide App. C			100	410	II (2) C						EPA-C TLV-A4
1,2-Dichloroethylene, all isomers (Acetylene dichloride) 156-59-2; 156-60-5; 540-59-0	200	793			200	790			200	790			200	800	II (2)						EPA-II* *CAS: 156-59-2; 156-60-5

SUBSTANCE CAS#	ACGIH® TLVs® TWA ppm	mg/m³	STEL/CEIL(C) ppm	mg/m³	OSHA PELs TWA ppm	mg/m³	STEL/CEIL(C) ppm	mg/m³	NIOSH RELs TWA ppm	mg/m³	STEL/CEIL(C) ppm	mg/m³	DFG MAKs TWA ppm	mg/m³	PEAK/CEIL(C) ppm	mg/m³	AIHA WEELs TWA ppm	mg/m³	STEL/CEIL(C) ppm	mg/m³	CARCINOGENICITY CATEGORY
Dichloroethyl ether (bis[2-Chloroethyl]ether) 111-44-4	5	29	10	58			C 15	C 90	5	30	10	60	10	59	I (1)						EPA-B2 IARC-3 NIOSH-Ca TLV-A4
		Skin				Skin			Skin See Pocket Guide App. A				Skin								
1,1-Dichloro-1-fluoro-ethane 1717-00-6																	500	2370	3000*	14,220*	
																			*5 min	
Dichlorofluoromethane (FC-21) 75-43-4	10	42			1000	4200			10	40			10	43	II (2)						
Dichloromethane (Methylene chloride) 75-09-2	50	174			25			125					50	180	II (2)						EPA-L NTP-R IARC-2B OSHA-Ca MAK-5 TLV-A3 NIOSH-Ca
		BEI			See 29 CFR 1910.1052				See Pocket Guide App. A				Skin; B								
1,2-Dichloromethoxy-ethane 41683-62-9																					MAK-3B
													Skin								
3,4-Dichloronitroben-zene 99-54-7																	1				MAK-3B
													Skin								
1,1-Dichloro-1-nitro-ethane 594-72-9	2	12					C 10	C 60	2	10											
2,4-Dichlorophenol 120-83-2																	1				IARC-2B
																Skin; Q				
4-(2,4-Dichlorophenoxy)-benzenamine 14861-17-7																					MAK-3B
													Skin								

SUBSTANCE / CAS#	ACGIH® TLVs® TWA ppm	mg/m³	STEL/CEIL(C) ppm	mg/m³	OSHA PELs TWA ppm	mg/m³	STEL/CEIL(C) ppm	mg/m³	NIOSH RELs TWA ppm	mg/m³	STEL/CEIL(C) ppm	mg/m³	DFG MAKs TWA ppm	mg/m³	PEAK/CEIL(C) ppm	mg/m³	AIHA WEELs TWA ppm	mg/m³	STEL/CEIL(C) ppm	mg/m³	CARCINOGENICITY CATEGORY
2,6-Dichloro-p-phenyl-enediamine 609-20-1																					IARC-3
1,3-Dichloro-2-propanol 96-23-1														Skin							IARC-2B MAK-2
1,3-Dichloropropene 542-75-6	1	4.5							1	5				cis- and trans-isomers							EPA-B2; K NIOSH-Ca IARC-2B* NTP-R* MAK-2 TLV-A3 *technical grade
	Skin								Skin See Pocket Guide App. A				Skin; Sh								
2,2-Dichloropropionic acid 75-99-0	5 I								1	6											TLV-A4
Dichlorotetrafluoroethane (1,2-Dichloro-1,1,2,2-tetrafluoroethane) 76-14-2	1000	6990			1000	7000			1000	7000			1000	7100	II (8)						TLV-A4
														D							
2,2-Dichloro-1,1,1-tri-fluoroethane (FC-123) 306-83-2																	50				MAK-3B
Dichlorvos (DDVP) 62-73-7	0.1 IFV				1				1				0.11	1	II (2)						EPA-B2 IARC-2B TLV-A4
	Skin; DSEN; BEI_A				Skin				Skin				Skin; C								
Dicofol 115-32-2																					IARC-3
Dicrotophos 141-66-2	0.05 IFV								0.25												TLV-A4
	Skin; BEI_A								Skin												

SUBSTANCE CAS#	ACGIH® TLVs® TWA ppm	mg/m³	STEL/CEIL(C) ppm	mg/m³	OSHA PELs TWA ppm	mg/m³	STEL/CEIL(C) ppm	mg/m³	NIOSH RELs TWA ppm	mg/m³	STEL/CEIL(C) ppm	mg/m³	DFG MAKs TWA ppm	mg/m³	PEAK/CEIL(C) ppm	mg/m³	AIHA WEELs TWA ppm	mg/m³	STEL/CEIL(C) ppm	mg/m³	CARCINOGENICITY CATEGORY
Dicyclohexylamine 101-83-7														Skin							
1,3-Dicyclohexyl-carbodiimide 538-75-0														Sh							
Dicyclopentadiene 77-73-6	5	27							5	30			0.5	2.7	I (1) D						
Dicyclopentadienyl iron, as Fe (Ferrocene) 102-54-5		10			15*; 5**				10*; 5**												
					*Total dust **Respirable fraction				*Total dust **Respirable fraction												
Didanosine 69655-05-6																					IARC-3
Di(tert-dodecyl) penta-sulfide 31565-23-8													100 I		II (2) C						
Di(tert-dodecyl) poly-sulfides 68425-15-0; 68583-56-2													100 I		II (2) C						
Dieldrin 60-57-1	0.1 IFV				0.25				0.25 Skin See Pocket Guide App. A				0.25 I Skin		II (8)						EPA-B2 IARC-3 NIOSH-Ca TLV-A3
		Skin				Skin															
Diepoxybutane 1464-53-5														2							IARC-1 NTP-R

SUBSTANCE / CAS#	ACGIH® TLVs® TWA ppm	TWA mg/m³	STEL/CEIL(C) ppm	STEL/CEIL(C) mg/m³	OSHA PELs TWA ppm	TWA mg/m³	STEL/CEIL(C) ppm	STEL/CEIL(C) mg/m³	NIOSH RELs TWA ppm	TWA mg/m³	STEL/CEIL(C) ppm	STEL/CEIL(C) mg/m³	DFG MAKs TWA ppm	TWA mg/m³	PEAK/CEIL(C) ppm	PEAK/CEIL(C) mg/m³	AIHA WEELs TWA ppm	TWA mg/m³	STEL/CEIL(C) ppm	STEL/CEIL(C) mg/m³	CARCINOGENICITY CATEGORY
Diesel engine emissions									*See* Pocket Guide App. A												EPA-L NTP-R IARC-1 MAK-2 NIOSH-Ca
Diesel fuel 68334-30-5; 68476-30-2; 68476-34-6; 77650-28-3	100 **IFV** as total hydrocarbons Skin																				IARC-2B TLV-A3
Diesel fuels, distillate (light)																					IARC-3
Diethanolamine 111-42-2	0.2 **IFV**	1 **IFV** Skin							3	15			1 **I** Skin; Sh; C		I (1)						IARC-2B MAK-3B TLV-A3
Diethylamine 109-89-7	5	15 Skin	15	45	25	75			10	30	25	75	2	6.1 Skin; D	I (2) C 5 C 15						TLV-A4
2-Diethylaminoethanol 100-37-8	2	9.6 Skin			10	50 Skin			10	50 Skin			5	24 Skin; C	I (1)						
Diethylbenzene, mixed isomers 25340-17-4																	5				
Diethylcarbamoyl chloride 88-10-8																					MAK-3B
Diethylene glycol 111-46-6													10	44	II (4) C		10				

SUBSTANCE / CAS#	ACGIH® TLVs® TWA ppm	mg/m³	STEL/CEIL(C) ppm	mg/m³	OSHA PELs TWA ppm	mg/m³	STEL/CEIL(C) ppm	mg/m³	NIOSH RELs TWA ppm	mg/m³	STEL/CEIL(C) ppm	mg/m³	DFG MAKs TWA ppm	mg/m³	PEAK/CEIL(C) ppm	mg/m³	AIHA WEELs TWA ppm	mg/m³	STEL/CEIL(C) ppm	mg/m³	CARCINOGENICITY CATEGORY
Diethylene glycol diacrylate 4074-88-8														Sh							
Diethylene glycol dimethacrylate 2358-84-1														Sh							
Diethylene glycol dimethyl ether 111-96-6													5	28	II (8) Skin; B						
Diethylene glycol dinitrate 693-21-0														Skin							EPA-D
Diethylene glycol monobutyl ether 112-34-5	10*	67.5* *IFV							10*	67				I (1.5) *sum of the concentrations of DGBE and its acetate in air C							
Diethylene triamine 111-40-0	1	4.2 Skin							1	4 Skin				Sh							
Di(2-ethylhexyl)adipate 103-23-1																					EPA-C IARC-3
Di(2-ethylhexyl)phthalate (DEHP) 117-81-7		5				5				5		10 See Pocket Guide App. A		2 I	II (2) Skin; C						EPA-B2 NIOSH-Ca IARC-2B NTP-R MAK-4 TLV-A3
1,2-Diethylhydrazine 1615-80-1																					IARC-2B

SUBSTANCE / CAS#	ACGIH® TLVs® TWA ppm	TWA mg/m³	STEL/CEIL(C) ppm	STEL/CEIL(C) mg/m³	OSHA PELs TWA ppm	TWA mg/m³	STEL/CEIL(C) ppm	STEL/CEIL(C) mg/m³	NIOSH RELs TWA ppm	TWA mg/m³	STEL/CEIL(C) ppm	STEL/CEIL(C) mg/m³	DFG MAKs TWA ppm	TWA mg/m³	PEAK/CEIL(C) ppm	PEAK/CEIL(C) mg/m³	AIHA WEELs TWA ppm	TWA mg/m³	STEL/CEIL(C) ppm	STEL/CEIL(C) mg/m³	CARCINOGENICITY CATEGORY
N,N-Diethylhydroxyla-mine 3710-84-7	2	7.3																			
Diethyl ketone 96-22-0	200	705	300	1057					200	705											
Diethyl-p-nitrophenyl-phosphate 311-45-5																					EPA-D
Diethyl phthalate 84-66-2		5								5											EPA-D TLV-A4
Diethylstilboestrol 56-53-1																					IARC-1 NTP-K
Diethyl sulfate 64-67-5															Skin; 2						IARC-2A MAK-2 NTP-R
N,N′-Diethylthiourea 105-55-5																					IARC-3
Difluorodibromo-methane 75-61-6	100	858			100	860			100	860											
1,1-Difluoroethane 75-37-6																	1000				

SUBSTANCE / CAS#	ACGIH® TLVs® TWA ppm	mg/m³	STEL/CEIL(C) ppm	mg/m³	OSHA PELs TWA ppm	mg/m³	STEL/CEIL(C) ppm	mg/m³	NIOSH RELs TWA ppm	mg/m³	STEL/CEIL(C) ppm	mg/m³	DFG MAKs TWA ppm	mg/m³	PEAK/CEIL(C) ppm	mg/m³	AIHA WEELs TWA ppm	mg/m³	STEL/CEIL(C) ppm	mg/m³	CARCINOGENICITY CATEGORY
Difluoromethane 75-10-5																	1000				
Diglycidyl ether (DGE) 2238-07-5	0.01	0.05					C 0.5	C 2.8	0.1	0.5			Skin								MAK-3B NIOSH-Ca TLV-A4
									See Pocket Guide App. A												
Diglycidyl hexanediol 16096-31-4													Sh								
Diglycidyl resorcinol ether 101-90-6													Skin; Sh								IARC-2B MAK-2 NTP-R
Digoxin 20830-75-5																					IARC-2B
1,2-Dihydroacean-thrylene 641-48-5																					IARC-3
Dihydrosafrole 94-58-6																					IARC-2B
Dihydroxymethyl-furatrizine 794-93-4																					IARC-3
Diisobutylene 107-39-1																	75				

SUBSTANCE CAS#	ACGIH® TLVs® TWA ppm	ACGIH® TLVs® TWA mg/m³	ACGIH® TLVs® STEL/CEIL(C) ppm	ACGIH® TLVs® STEL/CEIL(C) mg/m³	OSHA PELs TWA ppm	OSHA PELs TWA mg/m³	OSHA PELs STEL/CEIL(C) ppm	OSHA PELs STEL/CEIL(C) mg/m³	NIOSH RELs TWA ppm	NIOSH RELs TWA mg/m³	NIOSH RELs STEL/CEIL(C) ppm	NIOSH RELs STEL/CEIL(C) mg/m³	DFG MAKs TWA ppm	DFG MAKs TWA mg/m³	DFG MAKs PEAK/CEIL(C) ppm	DFG MAKs PEAK/CEIL(C) mg/m³	AIHA WEELs TWA ppm	AIHA WEELs TWA mg/m³	AIHA WEELs STEL/CEIL(C) ppm	AIHA WEELs STEL/CEIL(C) mg/m³	CARCINOGENICITY CATEGORY
Diisobutyl ketone (2,6-Dimethyl-4-heptanone) 108-83-8	25	145			50	290			25	150											
Diisodecyl phthalate 26761-40-0																					MAK-3B
Diisopropylamine 108-18-9	5	21 Skin			5	20 Skin			5	20 Skin											
Diisopropyl methyl-phosphonate (DIMP) 1445-75-6																					EPA-D
Diisopropyl sulfate 2973-10-6																					IARC-2B
Dimethipin 55290-64-7																					EPA-C
Dimethoxane 828-00-2																					IARC-3
3,3′-Dimethoxybenzidine (o-Dianisidine-based dyes) 119-90-4									*See* Pocket Guide Apps. A and C												IARC-2B MAK-2 NIOSH-Ca NTP-R
3,3′-Dimethoxybenzidine [119-90-4], **dyes metabolized to this compound**																					NTP-R

SUBSTANCE / CAS#	ACGIH® TLVs® TWA ppm	mg/m³	STEL/CEIL(C) ppm	mg/m³	OSHA PELs TWA ppm	mg/m³	STEL/CEIL(C) ppm	mg/m³	NIOSH RELs TWA ppm	mg/m³	STEL/CEIL(C) ppm	mg/m³	DFG MAKs TWA ppm	mg/m³	PEAK/CEIL(C)	AIHA WEELs TWA ppm	mg/m³	STEL/CEIL(C) ppm	mg/m³	CARCINOGENICITY CATEGORY
3,3′-Dimethoxybenzidine-4,4′-diisocyanate 91-93-0																				IARC-3
2,5-Dimethoxy-4-chloroaniline 6358-64-1														Skin						MAK-3B
N,N-Dimethylacetamide 127-19-5	10 NIC-A3	36 Skin; BEI			10	35 Skin			10	35 Skin			10	36 Skin; C	II (2)					(TLV-A4)
Dimethylamine 124-40-3	5	9.2 DSEN	15	27.6	10	18			10	18			2	3.7 D	I (2)	1			3	TLV-A4
4-Dimethylaminoazo-benzene 60-11-7							See 29 CFR 1910.1003				See Pocket Guide App. A									IARC-2B NIOSH-Ca NTP-R OSHA-Ca
p-Dimethylaminoazo-benzenediazo sodium sulfonate 140-56-7																				IARC-3
bis(2-Dimethylamino-ethyl) ether (DMAEE) 3033-62-3	0.05	0.33 Skin	0.15	0.98						minimize exposure See also NIAX® Catalyst ESN See Pocket Guide App. C										
N,N′-(Dimethylamino)ethyl methacrylate 2867-47-2														Sh						

SUBSTANCE / CAS#	ACGIH® TLVs® TWA ppm	mg/m³	STEL/CEIL(C) ppm	mg/m³	OSHA PELs TWA ppm	mg/m³	STEL/CEIL(C) ppm	mg/m³	NIOSH RELs TWA ppm	mg/m³	STEL/CEIL(C) ppm	mg/m³	DFG MAKs TWA ppm	mg/m³	PEAK/CEIL(C) ppm	mg/m³	AIHA WEELs TWA ppm	mg/m³	STEL/CEIL(C) ppm	mg/m³	CARCINOGENICITY CATEGORY
trans-2-[(Dimethyl-amino)methylimino]-5-[2-(5-nitro-2-furyl)-vinyl]-1,3,4-oxadiazole 25962-77-0																					IARC-2B
Dimethylaminopro-pionitrile 1738-25-6									minimize exposure *See* also NIAX® Catalyst ESN *See* Pocket Guide App. C												
4,4'-Dimethylangelicin plus ultraviolet A radiation 22975-76-4																					IARC-3
4,5'-Dimethylangelicin plus ultraviolet A radiation 4063-41-6																					IARC-3
Dimethylaniline (N,N-Dimethylaniline) 121-69-7	5	25	10	50	5	25			5	25	10	50	5	25	II (2)						IARC-3 MAK-3B TLV-A4
	Skin; BEI$_M$				Skin				Skin				Skin; D								
N-(1,3-Dimethyl butyl)-N'-phenyl-p-phenylene-diamine 793-24-8													2 I		II (2)						
													Sh; C								
Dimethyl carbamoyl chloride 79-44-7	0.005	0.02																			IARC-2A TLV-A2 MAK-2 NIOSH-Ca NTP-R
	Skin								*See* Pocket Guide App. A				Skin								
Dimethyldichlorosilane 75-78-5																			C 2		

SUBSTANCE / CAS#	ACGIH® TLVs® TWA ppm	TWA mg/m³	STEL/CEIL(C) ppm	STEL/CEIL(C) mg/m³	OSHA PELs TWA ppm	TWA mg/m³	STEL/CEIL(C) ppm	STEL/CEIL(C) mg/m³	NIOSH RELs TWA ppm	TWA mg/m³	STEL/CEIL(C) ppm	STEL/CEIL(C) mg/m³	DFG MAKs TWA ppm	TWA mg/m³	PEAK/CEIL(C) ppm	PEAK/CEIL(C) mg/m³	AIHA WEELs TWA ppm	TWA mg/m³	STEL/CEIL(C) ppm	STEL/CEIL(C) mg/m³	CARCINOGENICITY CATEGORY
Dimethyl disulfide 624-92-0	0.5	2 Skin																			
Dimethyl ether 115-10-6													1000	1900	II (8) D		1000				
Dimethylethoxysilane 14857-34-2	0.5	2.1	1.5	6.4																	
N,N-Dimethylethyl-amine 598-56-1													2	6.1	I (2) D						
Dimethylformamide 68-12-2	(10) NIC-5 NIC-A3	(30) NIC-15 Skin; BEI			˙0	30 Skin			10	30 Skin			5	15 Skin; B	II (2)						IARC-3 MAK-4 (TLV-A4)
1,1-Dimethylhydrazine 57-14-7	0.01	0.025 Skin			0.5	1 Skin					C 0.06* See Pocket Guide App. A	C 0.15* *120-min		Skin; Sh							IARC-2B NTP-R MAK-2 TLV-A3 NIOSH-Ca
1,2-Dimethylhydrazine 540-73-8														Skin; Sh							IARC-2A MAK-2
Dimethyl hydrogen phosphite 868-85-9																					IARC-3 MAK-3B
N,N-Dimethylisopropyl-amine 996-35-0													1	3.6	I (2) D						

SUBSTANCE / CAS#	ACGIH® TLVs® TWA ppm	mg/m³	STEL/CEIL(C) ppm	mg/m³	OSHA PELs TWA ppm	mg/m³	STEL/CEIL(C) ppm	mg/m³	NIOSH RELs TWA ppm	mg/m³	STEL/CEIL(C) ppm	mg/m³	DFG MAKs TWA ppm	mg/m³	PEAK/CEIL(C) ppm	mg/m³	AIHA WEELs TWA ppm	mg/m³	STEL/CEIL(C) ppm	mg/m³	CARCINOGENICITY CATEGORY
Dimethyloldihydroxy-ethylene urea 1854-26-8														Sh							
1,3-Dimethylol-5,5-di-methyl hydantoin 6440-58-0														Sh							
1,4-Dimethylphenan-threne 22349-59-3																					IARC-3
Dimethylphthalate 131-11-3		5				5				5											EPA-D
N,N-Dimethyl-p-tolui-dine 99-97-8																		0.5			
Dimethylsulfamoyl chloride 13360-57-1														Skin							MAK-2
Dimethyl sulfate 77-78-1	0.1	0.52			1	5			0.1	0.5											EPA-B2 NTP-R IARC-2A TLV-A3 MAK-2 NIOSH-Ca
		Skin				Skin			Skin See Pocket Guide App. A					Skin							
Dimethyl sulfide 75-18-3	10	25																			
Dimethyl sulfoxide 67-68-5													50	160	I (2)		250				
														Skin; B							

SUBSTANCE / CAS#	ACGIH® TLVs® TWA ppm	mg/m³	STEL/CEIL(C) ppm	mg/m³	OSHA PELs TWA ppm	mg/m³	STEL/CEIL(C) ppm	mg/m³	NIOSH RELs TWA ppm	mg/m³	STEL/CEIL(C) ppm	mg/m³	DFG MAKs TWA ppm	mg/m³	PEAK/CEIL(C) ppm	mg/m³	AIHA WEELs TWA ppm	mg/m³	STEL/CEIL(C) ppm	mg/m³	CARCINOGENICITY CATEGORY
Dimethyl terephthalate 120-61-6																		5* *Total dust			
Dimethyltin compounds, as Sn,** except Dimethyltin bis(isooctylmercaptoacetate) [26636-01-1], Dimethyltin bis(2-ethylhexylmercaptoacetate) [57583-35-4], bis[Dimethyltin(isooctylmercaptoacetate)]sulfide, and bis[Dimethyltin(2-mercaptoethyloleate)] sulfide													0.004	0.02*	I (1) *can also occur as vapor and aerosol **for methyltin cmpds whose organic ligands are already designated "Sa" or "Sh", these designations also apply C						
Dimethyltin bis(isooctylmercaptoacetate) [26636-01-1], Dimethyltin bis(2-ethylhexylmercaptoacetate) [57583-35-4], bis[Dimethyltin(isooctylmercaptoacetate)]sulfide, and bis[Dimethyltin(2-mercaptoethyloleate)]sulfide**													0.01	0.05*	II (2) *can also occur as vapor and aerosol **for methyltin cmpds whose organic ligands are already designated "Sa" or "Sh", these designations also apply C						
2,4-Dinitroanisole 119-27-7																	0.01*	0.1* *OARS WEEL			

SUBSTANCE / CAS#	ACGIH® TLVs® TWA ppm	ACGIH® TLVs® TWA mg/m³	ACGIH® TLVs® STEL/CEIL(C) ppm	ACGIH® TLVs® STEL/CEIL(C) mg/m³	OSHA PELs TWA ppm	OSHA PELs TWA mg/m³	OSHA PELs STEL/CEIL(C) ppm	OSHA PELs STEL/CEIL(C) mg/m³	NIOSH RELs TWA ppm	NIOSH RELs TWA mg/m³	NIOSH RELs STEL/CEIL(C) ppm	NIOSH RELs STEL/CEIL(C) mg/m³	DFG MAKs TWA ppm	DFG MAKs TWA mg/m³	DFG MAKs PEAK/CEIL(C) ppm	DFG MAKs PEAK/CEIL(C) mg/m³	AIHA WEELs TWA ppm	AIHA WEELs TWA mg/m³	AIHA WEELs STEL/CEIL(C) ppm	AIHA WEELs STEL/CEIL(C) mg/m³	CARCINOGENICITY CATEGORY
Dinitrobenzene, all isomers 99-65-0; 100-25-4; 528-29-0; 25154-54-5	0.15	1				1				1											EPA-D* MAK-3B *o-, m-isomers
	Skin; BEI$_M$				Skin				Skin				Skin								
3,7-Dinitrofluoranthene 105735-71-5																					IARC-2B
3,9-Dinitrofluoranthene 22506-53-2																					IARC-2B
4,6-Dinitro-o-cresol 534-52-1		0.2				0.2				0.2											
	Skin				Skin				Skin				Skin								
Dinitronaphthalene, all isomers 27478-34-8																					MAK-3B
1,3-Dinitropyrene 75321-20-9																					IARC-2B
1,6-Dinitropyrene 42397-64-8																					IARC-2B NTP-R
1,8-Dinitropyrene 42397-65-9																					IARC-2B NTP-R
3,5-Dinitro-o-toluamide (Dinitolmide) 148-01-6		1								5											TLV-A4

SUBSTANCE / CAS#	ACGIH® TLVs® TWA ppm	mg/m³	STEL/CEIL(C) ppm	mg/m³	OSHA PELs TWA ppm	mg/m³	STEL/CEIL(C) ppm	mg/m³	NIOSH RELs TWA ppm	mg/m³	STEL/CEIL(C) ppm	mg/m³	DFG MAKs TWA ppm	mg/m³	PEAK/CEIL(C) ppm	mg/m³	AIHA WEELs TWA ppm	mg/m³	STEL/CEIL(C) ppm	mg/m³	CARCINOGENICITY CATEGORY
Dinitrosopentamethylenetetramine 101-25-7																					IARC-3
Dinitrotoluene 25321-14-6		0.2				1.5				1.5 Skin *See* Pocket Guide App. A				mixture of isomers Skin							EPA-B2* NIOSH-Ca MAK-2* TLV-A3 *mixture of isomers
	Skin; BEI_M				Skin																
2,4-Dinitrotoluene 121-14-2																					IARC-2B
2,6-Dinitrotoluene 606-20-2																					IARC-2B
3,5-Dinitrotoluene 618-85-9																					IARC-3
Dinoseb 88-85-7																					EPA-D
Di-n-octyltin compounds, as Sn													0.002*	0.0098 I Skin**; B	II (2)						MAK-4
													*can also be found as vapor **for n-octyltin cmpds whose organic ligands are already designated "Sa" or "Sh", these designations also apply								
1,4-Dioxane (Diethylene dioxide) 123-91-1	20	72			100	360				C 1* *See* Pocket Guide App. A	C 3.6* *30-min		20	73 Skin; C	I (2)						EPA-L NTP-R IARC-2B TLV-A3 MAK-4 NIOSH-Ca
	Skin				Skin																

SUBSTANCE CAS#	ACGIH® TLVs® TWA ppm	ACGIH® TLVs® TWA mg/m³	ACGIH® TLVs® STEL/CEIL(C) ppm	ACGIH® TLVs® STEL/CEIL(C) mg/m³	OSHA PELs TWA ppm	OSHA PELs TWA mg/m³	OSHA PELs STEL/CEIL(C) ppm	OSHA PELs STEL/CEIL(C) mg/m³	NIOSH RELs TWA ppm	NIOSH RELs TWA mg/m³	NIOSH RELs STEL/CEIL(C) ppm	NIOSH RELs STEL/CEIL(C) mg/m³	DFG MAKs TWA ppm	DFG MAKs TWA mg/m³	DFG MAKs PEAK/CEIL(C) ppm	DFG MAKs PEAK/CEIL(C) mg/m³	AIHA WEELs TWA ppm	AIHA WEELs TWA mg/m³	AIHA WEELs STEL/CEIL(C) ppm	AIHA WEELs STEL/CEIL(C) mg/m³	CARCINOGENICITY CATEGORY
Dioxathion 78-34-2		0.1 **IFV** Skin; BEI_A								0.2 Skin											TLV-A4
1,3-Dioxolane 646-06-0	20	61											100	310 Skin; B	II (2)						
Dipentamethylene-thiuram disulfide 94-37-1														Sh							
Diphenylamine 122-39-4		10								10				5 **I** Skin; C	II (2)						MAK-3B TLV-A4
2,4'-Diphenyldiamine 492-17-1																					IARC-3
N,N-Diphenyl-p-phenylene diamine 74-31-7														Sh							
Dipropylene glycol 25265-71-8													100 **I*** *can also occur as vapor and aerosol C		II (2)						
Di(2-propylheptyl)-phthalate 53306-54-0																					MAK-3B
Dipropyl ketone 123-19-3	50	233							50	235											

SUBSTANCE / CAS#	ACGIH® TLVs® TWA ppm	TWA mg/m³	STEL/CEIL(C) ppm	STEL/CEIL(C) mg/m³	OSHA PELs TWA ppm	TWA mg/m³	STEL/CEIL(C) ppm	STEL/CEIL(C) mg/m³	NIOSH RELs TWA ppm	TWA mg/m³	STEL/CEIL(C) ppm	STEL/CEIL(C) mg/m³	DFG MAKs TWA ppm	TWA mg/m³	PEAK/CEIL(C) ppm	PEAK/CEIL(C) mg/m³	AIHA WEELs TWA ppm	TWA mg/m³	STEL/CEIL(C) ppm	STEL/CEIL(C) mg/m³	CARCINOGENICITY CATEGORY
Diquat 85-00-7; 2764-72-9; 6385-62-2		0.5 **I** 0.1 **R** Skin								0.5											TLV-A4
Disperse Blue 1 2475-45-8																					IARC-2B NTP-R
Disperse Blue 106/124 15141-18-1; 61951-51-7; 68516-81-4														Sh							
Disperse Orange 3 730-40-5														Sh							
Disperse Red 1 2872-52-8														Sh							
Disperse Red 17 3179-89-3														Sh							
Disperse Yellow 3 2832-40-8														Sh							IARC-3
Disulfiram 97-77-8		2								2* *precautions should be taken to avoid concurrent exposure to Ethylene dibromide				2 **I** Sh; D	II (8)						IARC-3 TLV-A4
Disulfoton 298-04-4		0.05 **IFV** Skin; BEI_A								0.1 Skin											TLV-A4

SUBSTANCE / CAS#	ACGIH® TLVs® TWA ppm	mg/m³	STEL/CEIL(C) ppm	mg/m³	OSHA PELs TWA ppm	mg/m³	STEL/CEIL(C) ppm	mg/m³	NIOSH RELs TWA ppm	mg/m³	STEL/CEIL(C) ppm	mg/m³	DFG MAKs TWA ppm	mg/m³	PEAK/CEIL(C) ppm	mg/m³	AIHA WEELs TWA ppm	mg/m³	STEL/CEIL(C) ppm	mg/m³	CARCINOGENICITY CATEGORY
1,4-Dithiane 505-29-3																					EPA-D
2,2'-Dithiobis(N-methyl-benzamide) 2527-58-4													Sh								
Dithranol 1143-38-0																					IARC-3
Diuron 330-54-1		10								10											TLV-A4
Divinylbenzene 1321-74-0	10	53							10	50											
Dodecyl mercaptan 112-55-0	0.1	0.8 DSEN									C 0.5*	C 4.1* *15-min									
Dowtherm® Q 612-00-0; 68987-42-8																	1				
Doxefazepam 40762-15-0																					IARC-3
Doxylamine succinate 562-10-7																					IARC-3

SUBSTANCE CAS#	ACGIH® TLVs® TWA ppm	mg/m³	STEL/CEIL(C) ppm	mg/m³	OSHA PELs TWA ppm	mg/m³	STEL/CEIL(C) ppm	mg/m³	NIOSH RELs TWA ppm	mg/m³	STEL/CEIL(C) ppm	mg/m³	DFG MAKs TWA ppm	mg/m³	PEAK/CEIL(C) ppm	mg/m³	AIHA WEELs TWA ppm	mg/m³	STEL/CEIL(C) ppm	mg/m³	CARCINOGENICITY CATEGORY
Droloxifene 82413-20-5																					IARC-3
Dulcin 150-69-6																					IARC-3
Dust, general threshold limit value														4 I							
Dust, general threshold limit value, biopersistent granular dusts, excluding ultrafine particles														0.3 R	II (8) for dusts with a density of 1 g/cm³ C						MAK-4
Emery 1302-74-5	TLV® withdrawn; see Aluminum, metal and insoluble compounds				15*; 5** *Total dust **Respirable fraction									4 I 1.5 R D							
Endosulfan 115-29-7	0.006* *IFV	0.1* Skin							0.1 Skin												TLV-A4
Endrin 72-20-8		0.1 Skin				0.1 Skin			0.1 Skin					0.05 I Skin; C	II (8)						EPA-D IARC-3 TLV-A4
Enflurane 13838-16-9	75	566									C 2* *60-min; for exposure to waste anesthetic gases	C 15.1*		20	150 C	II (8)						TLV-A4

SUBSTANCE / CAS#	ACGIH® TLVs® TWA ppm	mg/m³	STEL/CEIL(C) ppm	mg/m³	OSHA PELs TWA ppm	mg/m³	STEL/CEIL(C) ppm	mg/m³	NIOSH RELs TWA ppm	mg/m³	STEL/CEIL(C) ppm	mg/m³	DFG MAKs TWA ppm	mg/m³	PEAK/CEIL(C) ppm	mg/m³	AIHA WEELs TWA ppm	mg/m³	STEL/CEIL(C) ppm	mg/m³	CARCINOGENICITY CATEGORY
Engine exhaust																					IARC-1*; 2B** *diesel **gasoline
Eosin 15086-94-9																					IARC-3
Epichlorohydrin (1-Chloro-2,3-epoxypropane) 106-89-8	0.5	1.9		Skin	5	19		Skin	*See* Pocket Guide App. A					Skin; Sh; 3B							EPA-B2 NTP-R IARC-2A TLV-A3 MAK-2 NIOSH-Ca
EPN (O-Ethyl O-[4-nitro-phenyl]phenylthiophos-phonate) 2104-64-5		0.1 I		Skin; BEI_A		0.5		Skin		0.5		Skin		0.5 I	II (2)		Skin				TLV-A4
1,2-Epoxybutane (1,2-Butylene oxide) 106-88-7															Skin		2				IARC-2B MAK-2
3,4-Epoxycyclohexyl-methyl-3,4-epoxycyclohex-ylcarboxylate 2386-87-0												Skin; Sh									MAK-3B
bis(2,3-Epoxycyclo-pentyl)ether 2386-90-5																					IARC-3
3,4-Epoxy-6-methyl-cyclohexylmethyl-3,4-epoxy-6-methyl-cyclohexane carbox-ylate 141-37-7																					IARC-3

SUBSTANCE CAS#	ACGIH® TLVs® TWA ppm	mg/m³	STEL/CEIL(C) ppm	mg/m³	OSHA PELs TWA ppm	mg/m³	STEL/CEIL(C) ppm	mg/m³	NIOSH RELs TWA ppm	mg/m³	STEL/CEIL(C) ppm	mg/m³	DFG MAKs TWA ppm	mg/m³	PEAK/CEIL(C) ppm	mg/m³	AIHA WEELs TWA ppm	mg/m³	STEL/CEIL(C) ppm	mg/m³	CARCINOGENICITY CATEGORY
9,10-Epoxystearic acid, cis-isomer 24560-98-3																					IARC-3
Erionite, fibrous dust 12510-42-8; 66733-21-9																					IARC-1* MAK-1 NTP-K* *CAS: 66733-21-9
Erythromycin 114-07-8																		3			
Estazolam 29975-16-4																					IARC-3
Estradiol mustard 22966-79-6																					IARC-3
Ethane 74-84-0	See Appendix F in TLVs® and BEIs® book (D, EX)																				
Ethanol (Ethyl alcohol) 64-17-5			1000	1880	1000	1900			1000	1900			500	960	II (2) C; 5						MAK-5 TLV-A3
Ethanolamine (2-Aminoethanol) 141-43-5	3	7.5	6	15	3	6			3	8	6	15	0.2	0.51*	I (1) Sh; C						*can also occur as vapor and aerosol
Ethidium bromide 1239-45-8															3B						MAK-3B

SUBSTANCE / CAS#	ACGIH® TLVs® TWA ppm	mg/m³	STEL/CEIL(C) ppm	mg/m³	OSHA PELs TWA ppm	mg/m³	STEL/CEIL(C) ppm	mg/m³	NIOSH RELs TWA ppm	mg/m³	STEL/CEIL(C) ppm	mg/m³	DFG MAKs TWA ppm	mg/m³	PEAK/CEIL(C) ppm	mg/m³	AIHA WEELs TWA ppm	mg/m³	STEL/CEIL(C) ppm	mg/m³	CARCINOGENICITY CATEGORY
Ethion 563-12-2		0.05 IFV Skin; BEI_A								0.4 Skin											TLV-A4
Ethionamide 536-33-4																					IARC-3
2-Ethoxyethanol (EGEE; Cellosolve) 110-80-5	5	18 Skin; BEI			200	740 Skin			0.5	1.8 Skin			2* *sum of the concentrations of EGEE and its acetate in air Skin; B	7.5	II (8)						
2-(2-Ethoxyethoxy)-ethanol (Diethylene glycol monoethyl ether) 111-90-0													50 I C		I (2)		25				
2-Ethoxyethyl acetate (EGEEA; Cellosolve acetate) 111-15-9	5	27 Skin; BEI			100	540 Skin			0.5	2.7 Skin			2* *sum of the concentrations of EGEE and its acetate in air Skin; B	11	II (8)						
1-Ethoxy-2-propanol 1569-02-4													50* *sum of the concentrations of CAS: 1569-02-4 and 54839-24-6 in air Skin; C	220	II (2)						
1-Ethoxy-2-propyl acetate 54839-24-6													50* *sum of the concentrations of CAS: 1569-02-4 and 54839-24-6 in air C	300	II (2)						
Ethyl acetate 141-78-6	400	1440			400	1400			400	1400			200 C	750	I (2)						
Ethyl acrylate (Acrylic acid, ethyl ester) 140-88-5	5	20	15	61	25	100 Skin				*See* Pocket Guide App. A			2 Skin; Sh; C	8.3	I (2)						IARC-2B NIOSH-Ca TLV-A4

SUBSTANCE / CAS#	ACGIH® TLVs® TWA ppm	mg/m³	STEL/CEIL(C) ppm	mg/m³	OSHA PELs TWA ppm	mg/m³	STEL/CEIL(C) ppm	mg/m³	NIOSH RELs TWA ppm	mg/m³	STEL/CEIL(C) ppm	mg/m³	DFG MAKs TWA ppm	mg/m³	PEAK/CEIL(C) ppm	mg/m³	AIHA WEELs TWA ppm	mg/m³	STEL/CEIL(C) ppm	mg/m³	CARCINOGENICITY CATEGORY
Ethylamine 75-04-7	5	9	15	28	10	18			10	18			5	9.4	I (2) C 10	19					
		Skin													D						
Ethyl amyl ketone (5-Methyl-3-heptanone) 541-85-5	10	52			25	130			25	130			10	53	I (2)						
															D						
Ethylbenzene 100-41-4	20	87			100	435			100	435	125	545	20	88	II (2)						EPA-D IARC-2B MAK-4 TLV-A3
		BEI												Skin; C							
Ethyl bromide (Bromoethane) 74-96-4	5	22			200	890															IARC-3 MAK-2 TLV-A3
		Skin												Skin							
Ethyl tert-butyl ether 637-92-3	25	105																			TLV-A4
Ethyl butyl ketone (3-Heptanone) 106-35-4	50	234	75	350	50	230			50	230			10	47	I (2)						
															D						
Ethyl chloride (Chloroethane) 75-00-3	100	264			1000	2600			handle with caution *See* Pocket Guide App. C												IARC-3 MAK-3B TLV-A3
		Skin												Skin							
Ethyl chloroformate (Chloroformic acid ethyl ester) 541-41-3																					MAK-3B
Ethyl cyanoacrylate (Ethyl 2-cyanoacrylate) 7085-85-0	(0.2)	(1)																			
	NIC-withdraw adopted TLV® and *Documentation; see* Cyanoacrylates, Ethyl and Methyl																				

SUBSTANCE / CAS#	ACGIH® TLVs® TWA ppm	mg/m³	STEL/CEIL(C) ppm	mg/m³	OSHA PELs TWA ppm	mg/m³	STEL/CEIL(C) ppm	mg/m³	NIOSH RELs TWA ppm	mg/m³	STEL/CEIL(C) ppm	mg/m³	DFG MAKs TWA ppm	mg/m³	PEAK/CEIL(C) ppm	mg/m³	AIHA WEELs TWA ppm	mg/m³	STEL/CEIL(C) ppm	mg/m³	CARCINOGENICITY CATEGORY
5-Ethyl-3,7-dioxa-1-aza-bicyclo[3.3.0]octane 7747-35-5														Sh							
Ethylene 74-85-1	200	230																			IARC-3 MAK-3B TLV-A4
Ethylene chlorohydrin (2-Chloroethanol) 107-07-3			C 1	C 3.3 Skin	5	16		Skin			C 1	C 3 Skin	1	3.3	II (1)	Skin; C					TLV-A4
Ethylenediamine (1,2-Diaminoethane) 107-15-3	10	25		Skin	10	25			10	25				Sah							EPA-D TLV-A4
Ethylene dibromide (1,2-Dibromoethane) 106-93-4				Skin	20		C 30; 50*	*5-min peak per 8-hr shift	0.045		C 0.13*	*15-min See Pocket Guide App. A		Skin							EPA-L NTP-R IARC-2A TLV-A3 MAK-2 NIOSH-Ca
Ethylene dichloride (1,2-Dichloroethane) 107-06-2	10	40			50		C 100; 200*	*5-min peak in any 3 hrs	1	4	2	8 See Pocket Guide Apps. A and C		Skin							EPA-B2 NIOSH-Ca IARC-2B NTP-R MAK-2 TLV-A4
Ethylene glycol 107-21-1	25*		50* *(V) **I; (H)	10**									10	26* *can also occur as vapor and aerosol	I (2)		I (2)				TLV-A4
Ethylene glycol dimeth-acrylate 97-90-5														Sh							

SUBSTANCE / CAS#	ACGIH® TLVs® TWA ppm	mg/m³	STEL/CEIL(C) ppm	mg/m³	OSHA PELs TWA ppm	mg/m³	STEL/CEIL(C) ppm	mg/m³	NIOSH RELs TWA ppm	mg/m³	STEL/CEIL(C) ppm	mg/m³	DFG MAKs TWA ppm	mg/m³	PEAK/CEIL(C) ppm	mg/m³	AIHA WEELs TWA ppm	mg/m³	STEL/CEIL(C) ppm	mg/m³	CARCINOGENICITY CATEGORY
Ethylene glycol dinitrate (EGDN) 628-96-6	0.05	0.31					C 0.2	C 1				0.1	0.01*	0.063**	II (1)						
													* applies for the sum of the concentrations of Ethylene glycol dinitrate and Nitroglycerin in air **can also occur as vapor and aerosol								
	Skin				Skin				Skin				Skin; C								
Ethylene glycol methacrylate (2-Hydroxyethyl methacrylate) 868-77-9													Sh								
Ethylene oxide (EtO) 75-21-8	1	1.8			1		5		< 0.1	0.18	C 5*	C 9* *10-min/day									EPA-CaH IARC-1 MAK-2 NIOSH-Ca NTP-K OSHA-Ca TLV-A2
				See 29 CFR 1910.1047(c)				See Pocket Guide App. A				Skin; 2								
Ethylene sulfide 420-12-2																					IARC-3
Ethylene thiourea 96-45-7									use encapsulated form See Pocket Guide App. A												IARC-3 MAK-3B NIOSH-Ca NTP-R
Ethyleneimine 151-56-4	0.05	0.09	0.1	0.18																	IARC-2B MAK-2 NIOSH-Ca OSHA-Ca TLV-A3
	Skin				See 29 CFR 1910.1003				See Pocket Guide App. A				Skin; 2								
Ethyl ether (Diethyl ether) 60-29-7	400	1210	500	1520	400	1200							400	1200	I (1)						
													D								
Ethyl-3-ethoxy-propionate 763-69-9													100	610	I (1)						
													Skin; C								

SUBSTANCE / CAS#	ACGIH® TLVs® TWA ppm	mg/m³	STEL/CEIL(C) ppm	mg/m³	OSHA PELs TWA ppm	mg/m³	STEL/CEIL(C) ppm	mg/m³	NIOSH RELs TWA ppm	mg/m³	STEL/CEIL(C) ppm	mg/m³	DFG MAKs TWA ppm	mg/m³	PEAK/CEIL(C) ppm	mg/m³	AIHA WEELs TWA ppm	mg/m³	STEL/CEIL(C) ppm	mg/m³	CARCINOGENICITY CATEGORY
Ethyl formate (Formic acid, ethyl ester) 109-94-4			100	303	100	300			100	300			100	310	I (1) Skin; C						TLV-A4
2-Ethylhexanoic acid 149-57-5		5 IFV																			
2-Ethylhexanol 104-76-7													10	54	I (1) C						
2-Ethylhexyl acetate 103-09-3													10	71	I (1) C						
2-Ethylhexyl acrylate (Acrylic acid, 2-ethylhexyl ester) 103-11-7													5	38	I (1) Sh; C						IARC-3
2-Ethylhexyl mercaptoacetate 7659-86-1															Sh						
N,N-bis(2-Ethylhexyl)-(1,2,4-triazole-1-yl)-methanamine 91273-04-0													can also occur as vapor and aerosol Sh								
Ethylidene norbornene 16219-75-3	2	10	4	20							C 5	C 25									
Ethyl isocyanate 109-90-0	0.02	0.06	0.06	0.17											Skin; DSEN						

SUBSTANCE / CAS#	ACGIH® TLVs® TWA ppm	TWA mg/m³	STEL/CEIL(C) ppm	STEL/CEIL(C) mg/m³	OSHA PELs TWA ppm	TWA mg/m³	STEL/CEIL(C) ppm	STEL/CEIL(C) mg/m³	NIOSH RELs TWA ppm	TWA mg/m³	STEL/CEIL(C) ppm	STEL/CEIL(C) mg/m³	DFG MAKs TWA ppm	TWA mg/m³	PEAK/CEIL(C) ppm	PEAK/CEIL(C) mg/m³	AIHA WEELs TWA ppm	TWA mg/m³	STEL/CEIL(C) ppm	STEL/CEIL(C) mg/m³	CARCINOGENICITY CATEGORY
Ethyl mercaptan (Ethanethiol) 75-08-1	0.5	1.3					C 10	C 25			C 0.5* *15 min	C 1.3*	0.5	1.3	II (2) D						
Ethyl methacrylate (Methacrylic acid, ethyl ester) 97-63-2															Sh						
Ethyl methanesulfonate 62-50-0																					IARC-2B NTP-R
Ethyl methyl ketoxime (2-Butanone oxime) 96-29-7															Skin; Sh		10		DSEN		MAK-2
N-Ethylmorpholine 100-74-3	5	24	Skin		20	94	Skin		5	23	Skin										
N-Ethylpyrrolidone 2687-91-4													5	23* *can also occur as vapor and aerosol	I (2) Skin; C						
Ethyl selenac 5456-28-0																					IARC-3
Ethyl silicate (Silicic acid tetraethyl ester) 78-10-4	10	85			100	850			10	85			10	86	I (1) D						
Ethyl telluric 20941-65-5																					IARC-3

OCCUPATIONAL EXPOSURE VALUES

SUBSTANCE / CAS#	ACGIH® TLVs® TWA ppm	mg/m³	STEL/CEIL(C) ppm	mg/m³	OSHA PELs TWA ppm	mg/m³	STEL/CEIL(C) ppm	mg/m³	NIOSH RELs TWA ppm	mg/m³	STEL/CEIL(C) ppm	mg/m³	DFG MAKs TWA ppm	mg/m³	PEAK/CEIL(C) ppm	mg/m³	AIHA WEELs TWA ppm	mg/m³	STEL/CEIL(C) ppm	mg/m³	CARCINOGENICITY CATEGORY
Ethyltin compounds 7440-31-5													Skin* *for Ethyltin cmpds whose organic ligands are already designated "Sa" or "Sh", these designations also apply								
Etoposide 33419-42-0																					IARC-1
Eugenol 97-53-0													Sh								IARC-3
Evans Blue 314-13-6																					IARC-3
Farnesol 4602-84-0													Sh								IARC-3
Fast Green FCF 2353-45-9																					IARC-3
Fenamiphos 22224-92-6	0.05 **IFV** Skin; BEI_A								0.1 Skin												TLV-A4
Fensulfothion 115-90-2	0.01 **IFV** Skin; BEI_A								0.1												TLV-A4

SUBSTANCE / CAS#	ACGIH® TLVs® TWA ppm	TWA mg/m³	STEL/CEIL(C) ppm	mg/m³	OSHA PELs TWA ppm	mg/m³	STEL/CEIL(C) ppm	mg/m³	NIOSH RELs TWA ppm	mg/m³	STEL/CEIL(C) ppm	mg/m³	DFG MAKs TWA ppm	mg/m³	PEAK/CEIL(C) ppm	mg/m³	AIHA WEELs TWA ppm	mg/m³	STEL/CEIL(C) ppm	mg/m³	CARCINOGENICITY CATEGORY
Fenthion 55-38-9		0.05 IFV Skin; BEI_A												0.2 I Skin	II (2)						TLV-A4
Fenvalerate 51630-58-1																					IARC-3
Ferbam 14484-64-1		5 I				15* *Total dust				10											IARC-3 TLV-A4
Ferrovanadium dust 12604-58-9	1			3		1			1*		3*										
									*also applies to Vanadium metal and Vanadium carbide												
Flour dust		0.5 I RSEN											Cereal (rye, wheat) flour dusts Sa								
Fludioxonil 131341-86-1		NIC-1 I NIC-A3																			
Fluometuron 2164-17-2																					IARC-3
Fluoranthene 206-44-0																					EPA-D IARC-3
Fluorene 86-73-7																					EPA-D IARC-3

SUBSTANCE / CAS#	ACGIH® TLVs® TWA ppm	mg/m³	STEL/CEIL(C) ppm	mg/m³	OSHA PELs TWA ppm	mg/m³	STEL/CEIL(C) ppm	mg/m³	NIOSH RELs TWA ppm	mg/m³	STEL/CEIL(C) ppm	mg/m³	DFG MAKs TWA ppm	mg/m³	PEAK/CEIL(C) ppm	mg/m³	AIHA WEELs TWA ppm	mg/m³	STEL/CEIL(C) ppm	mg/m³	CARCINOGENICITY CATEGORY
Fluorides, as F		2.5				2.5				2.5			1 I		II (4)						IARC-3* TLV-A4 *inorganic, used in drinking water
		BEI												Skin; C							
Fluorine 7782-41-4	1	1.6	2	3.1	0.1	0.2			0.1	0.2											
5-Fluorouracil 51-21-8																					IARC-3
Fluroxene 406-90-6											C 2*	C 10.3*									
									*60-min; for exposure to waste anesthetic gases												
Folpet 133-07-3		1 I																			EPA-B2 TLV-A3
		DSEN																			
Fomesafen 72178-02-0																					EPA-C
Fonofos 944-22-9		0.1 IFV								0.1											TLV-A4
		Skin; BEI_A								Skin											
Formaldehyde 50-00-0	0.1	0.12	0.3	0.37	0.75		2		0.016		C 0.1*		0.3	0.37	I (2)						EPA-B1 NTP-K IARC-1 OSHA-Ca MAK-4 TLV-A1 NIOSH-Ca
		DSEN; RSEN			*See* 29 CFR 1910.1048(c)				*15-min *See* Pocket Guide App. A				C 1		C 1.2						
													Sh; C; 5								
Formaldehyde condensation products with p-tert-butylphenol (low-molecular)														Sh							

SUBSTANCE / CAS#	ACGIH® TLVs® TWA ppm	mg/m³	STEL/CEIL(C) ppm	mg/m³	OSHA PELs TWA ppm	mg/m³	STEL/CEIL(C) ppm	mg/m³	NIOSH RELs TWA ppm	mg/m³	STEL/CEIL(C) ppm	mg/m³	DFG MAKs TWA ppm	mg/m³	PEAK/CEIL(C) ppm	mg/m³	AIHA WEELs TWA ppm	mg/m³	STEL/CEIL(C) ppm	mg/m³	CARCINOGENICITY CATEGORY
Formaldehyde condensation products with phenol (low-molecular)														Sh							
Formamide 75-12-7	10	18		Skin					10	15		Skin		Skin							
Formic acid 64-18-6	5	9.4	10	19	5	9			5	9			5	9.5	I (2)	C					
Fosetyl-al 39148-24-8																					EPA-C
Fuel oils, distillate (light)																					IARC-3
Fumonisin B₁ 116355-83-0																					IARC-2B
Furan 110-00-9													0.02	0.056	II (2)	Skin; D			(W)		IARC-2B MAK-4 NTP-R
Furazolidone 67-45-8																					IARC-3
Furfural 98-01-1	0.2	0.8		Skin; BEI	5	20		Skin						Skin							IARC-3 MAK-3B TLV-A3

SUBSTANCE / CAS#	ACGIH® TLVs® TWA ppm	mg/m³	STEL/CEIL(C) ppm	mg/m³	OSHA PELs TWA ppm	mg/m³	STEL/CEIL(C) ppm	mg/m³	NIOSH RELs TWA ppm	mg/m³	STEL/CEIL(C) ppm	mg/m³	DFG MAKs TWA ppm	mg/m³	PEAK/CEIL(C) ppm	mg/m³	AIHA WEELs TWA ppm	mg/m³	STEL/CEIL(C) ppm	mg/m³	CARCINOGENICITY CATEGORY
Furfuryl alcohol 98-00-0	0.2	0.8 Skin			50	200			10	40	15	60 Skin		Skin							MAK-3B TLV-A3
Furmecyclox 60568-05-0																					EPA-B2
Furosemide (Frusemide) 54-31-9																					IARC-3
Furothiazole 531-82-8																					IARC-2B
Furylfuramide 3688-53-7																					IARC-2B
Gallium arsenide 1303-00-0	0.0003 **R**									as As *See* Pocket Guide App. A	C 0.002*	*15-min									IARC-1* NIOSH-Ca TLV-A3 *as As
Gasoline 8006-61-9; 86290-81-5	300	890 bulk handling	500	1480						*See* Pocket Guide App. A											IARC-2B NIOSH-Ca TLV-A3
Gasoline, engine emissions																					IARC-2B
Gemfibrozil 25812-30-0																					IARC-3

SUBSTANCE / CAS#	ACGIH® TLVs® TWA ppm	TWA mg/m³	STEL/CEIL(C) ppm	STEL/CEIL(C) mg/m³	OSHA PELs TWA ppm	TWA mg/m³	STEL/CEIL(C) ppm	STEL/CEIL(C) mg/m³	NIOSH RELs TWA ppm	TWA mg/m³	STEL/CEIL(C) ppm	STEL/CEIL(C) mg/m³	DFG MAKs TWA ppm	TWA mg/m³	PEAK/CEIL(C) ppm	PEAK/CEIL(C) mg/m³	AIHA WEELs TWA ppm	TWA mg/m³	STEL/CEIL(C) ppm	STEL/CEIL(C) mg/m³	CARCINOGENICITY CATEGORY
Geraniol 106-24-1																					
													Sh								
Germanium tetra-hydride 7782-65-2	0.2	0.63							0.2	0.6											
Glutaraldehyde 111-30-8			C 0.05	C 0.2							C 0.2	C 0.8	0.05	0.21	I (2)						MAK-4 TLV-A4
activated or unactivated DSEN; RSEN									_See Pocket Guide App. C_						C 0.2	C 0.83					
													Sah; C								
Glycerin 56-81-5	Mist				15*; 5**								200 I		I (2)						
TLV® withdrawn due to insufficient data relevant to human occupational exposure					Mist *Total dust **Respirable fraction								C								
Glyceryl monothio-glycolate 30618-84-9													Sh								
Glycidaldehyde 765-34-4																					EPA-B2 IARC-2B
Glycidol (2,3-Epoxy-1-propanol) 556-52-5	2	6.1			50	150			25	75											IARC-2A MAK-2 NTP-R TLV-A3
													Skin; 3A								
Glycidyl methacrylate 106-91-2																	0.5				
													Sh				Skin; DSEN				
Glycidyl oleate 5431-33-4																					IARC-3

SUBSTANCE / CAS#	ACGIH® TLVs® TWA ppm	mg/m³	STEL/CEIL(C) ppm	mg/m³	OSHA PELs TWA ppm	mg/m³	STEL/CEIL(C) ppm	mg/m³	NIOSH RELs TWA ppm	mg/m³	STEL/CEIL(C) ppm	mg/m³	DFG MAKs TWA ppm	mg/m³	PEAK/CEIL(C) ppm	mg/m³	AIHA WEELs TWA ppm	mg/m³	STEL/CEIL(C) ppm	mg/m³	CARCINOGENICITY CATEGORY
Glycidyl stearate 7460-84-6																					IARC-3
Glycidyl trimethyl ammonium chloride 3033-77-0														Skin; Sh							MAK-2
Glycolonitrile 107-16-4											C 2*	C 5* *15-min									
Glyoxal 107-22-2	0.1 IFV	DSEN												Skin; Sh				0.1 DSEN; (H)			MAK-3B TLV-A4
Glyphosate 1071-83-6																					EPA-D IARC-2A
Gold [7440-57-5] and inorganic compounds														soluble compounds only Sh							
Grain dust (oat; wheat; barley)	4				10				4												
Graphite, natural 7782-42-5	2 R	all forms except Graphite fibers			15 mppcf* *based on impinger samples counted by light field techniques				2.5* *Respirable dust					1.5 R 4 I C							
Graphite, synthetic	2 R	all forms except Graphite fibers			15*; 5** *Total dust **Respirable fraction																

SUBSTANCE / CAS#	ACGIH® TLVs® TWA ppm	mg/m³	STEL/CEIL(C) ppm	mg/m³	OSHA PELs TWA ppm	mg/m³	STEL/CEIL(C) ppm	mg/m³	NIOSH RELs TWA ppm	mg/m³	STEL/CEIL(C) ppm	mg/m³	DFG MAKs TWA ppm	mg/m³	PEAK/CEIL(C) ppm	mg/m³	AIHA WEELs TWA ppm	mg/m³	STEL/CEIL(C) ppm	mg/m³	CARCINOGENICITY CATEGORY
Griseofulvin 126-07-8																					IARC-2B
Guinea Green B 4680-78-8																					IARC-3
Gyromitrin 16568-02-8																					IARC-3
Hafnium [7440-58-6] and compounds, as Hf		0.5				0.5				0.5											
Halloysite, fibrous dust 12298-43-0																					MAK-3B
Halothane 151-67-7	50	404									C 2*	C 16.2*	5	41	II (8)						TLV-A4
									*60-min; for exposure to waste anesthetic gases						B						
Hard metals containing Cobalt and Tungsten carbide		0.005* T *measured as Cobalt RSEN											Inhalable fraction Skin; Sah; 3A								MAK-1 TLV-A2
HC Blue No. 1 2784-94-3																					IARC-2B
HC Blue No. 2 33229-34-4																					IARC-3

SUBSTANCE / CAS#	ACGIH® TLVs® TWA ppm	TWA mg/m³	STEL/CEIL(C) ppm	STEL/CEIL(C) mg/m³	OSHA PELs TWA ppm	TWA mg/m³	STEL/CEIL(C) ppm	STEL/CEIL(C) mg/m³	NIOSH RELs TWA ppm	TWA mg/m³	STEL/CEIL(C) ppm	STEL/CEIL(C) mg/m³	DFG MAKs TWA ppm	TWA mg/m³	PEAK/CEIL(C) ppm	PEAK/CEIL(C) mg/m³	AIHA WEELs TWA ppm	TWA mg/m³	STEL/CEIL(C) ppm	STEL/CEIL(C) mg/m³	CARCINOGENICITY CATEGORY
HC Red No. 3 — 2871-01-4																					IARC-3
HC Yellow No. 4 — 59820-43-8																					IARC-3
Helium — 7440-59-7	*Documentation* withdrawn; *see* Appendix F in *TLVs® and BEIs®* book — Simple asphyxiant(D)																				
Hematite — 1317-60-8																					IARC-3; 1* *mining underground
Heptachlor — 76-44-8		0.05 Skin				0.5 Skin				0.5 Skin *See* Pocket Guide App. A				0.05 I Skin; D	II (8)						EPA-B2 TLV-A3 IARC-2B MAK-4 NIOSH-Ca
Heptachlor epoxide — 1024-57-3		0.05 Skin																			EPA-B2 IARC-2B TLV-A3
Heptane, isomers — 108-08-7; 142-82-5; 565-59-3; 589-34-4; 590-35-2; 591-76-4	400	1640	500	2050	500	2000 CAS: 142-82-5 only			85	350	C 440* *15-min CAS: 142-82-5 only	C 1800*	500	2100 CAS: 142-82-5 only D	I (1)						EPA-D* *CAS: 142-82-5 only
n-Heptyl mercaptan (1-Heptanethiol) — 1639-09-4											C 0.5* *15-min	C 2.7*									
Hexabromodiphenyl ether — 36483-60-0																					EPA-D

SUBSTANCE / CAS#	ACGIH® TLVs® TWA ppm	mg/m³	STEL/CEIL(C) ppm	mg/m³	OSHA PELs TWA ppm	mg/m³	STEL/CEIL(C) ppm	mg/m³	NIOSH RELs TWA ppm	mg/m³	STEL/CEIL(C) ppm	mg/m³	DFG MAKs TWA ppm	mg/m³	PEAK/CEIL(C) ppm	mg/m³	AIHA WEELs TWA ppm	mg/m³	STEL/CEIL(C) ppm	mg/m³	CARCINOGENICITY CATEGORY
2,2′,4,4′,5,5′-Hexabromodiphenyl ether (BDE-153) 68631-49-2																					EPA-I
Hexachlorobenzene (HCB) 118-74-1		0.002																			EPA-B2 TLV-A3 IARC-2B MAK-4 NTP-R
	Skin													Skin; D							
Hexachlorobutadiene 87-68-3	0.02	0.21							0.02	0.24			0.02	0.22	II (2)						EPA-C TLV-A3 IARC-3 MAK-4 NIOSH-Ca
	Skin								Skin See Pocket Guide App. A				Skin; C							
1,2,3,4,5,6-Hexachlorocyclohexane, mixture of α-HCH [319-84-6] and β-HCH [319-85-7] isomers														0.1* I	II (8)						EPA-B2*; C** MAK-4 *319-84-6 **319-85-7
													*(concentration α-HCH ÷ 5) + concentration β-HCH Skin; D							
Hexachlorocyclohexane technical (t-HCH) 608-73-1																					EPA-B2
α-Hexachlorocyclohexane 319-84-6														0.5 I	II (8)						MAK-4
													Skin; D							
Δ-Hexachlorocyclohexane 319-86-8																					EPA-D
ε-Hexachlorocyclohexane 6108-10-7																					EPA-D

SUBSTANCE CAS#	ACGIH® TLVs® TWA ppm	ACGIH® TLVs® TWA mg/m³	ACGIH® TLVs® STEL/CEIL(C) ppm	ACGIH® TLVs® STEL/CEIL(C) mg/m³	OSHA PELs TWA ppm	OSHA PELs TWA mg/m³	OSHA PELs STEL/CEIL(C) ppm	OSHA PELs STEL/CEIL(C) mg/m³	NIOSH RELs TWA ppm	NIOSH RELs TWA mg/m³	NIOSH RELs STEL/CEIL(C) ppm	NIOSH RELs STEL/CEIL(C) mg/m³	DFG MAKs TWA ppm	DFG MAKs TWA mg/m³	DFG MAKs PEAK/CEIL(C) ppm	DFG MAKs PEAK/CEIL(C) mg/m³	AIHA WEELs TWA ppm	AIHA WEELs TWA mg/m³	AIHA WEELs STEL/CEIL(C) ppm	AIHA WEELs STEL/CEIL(C) mg/m³	CARCINOGENICITY CATEGORY
Hexachlorocyclo-pentadiene 77-47-4	0.01	0.11							0.01	0.1					Skin						EPA-NL; E TLV-A4
Hexachlorodibenzo-p-dioxin, mixture (HxCDD) 19408-74-3; 57653-85-7																					EPA-B2
Hexachloroethane 67-72-1	1	9.7 Skin			1	10 Skin			1	10 Skin See Pocket Guide Apps. A and C			1	9.8	II (2)						EPA-L TLV-A3 IARC-2B NIOSH-Ca NTP-R
Hexachloronaphthalene 1335-87-1		0.2 Skin				0.2 Skin				0.2 Skin				Skin							
Hexachlorophene 70-30-4																					IARC-3
2,4-Hexadienal 142-83-6																					IARC-2B
1,4-Hexadiene 592-45-0																	10	34			
Hexafluoroacetone 684-16-2	0.1	0.68 Skin							0.1	0.7 Skin											
cis-1,1,1,4,4,4-Hexa-fluoro-2-butene 692-49-9																	500*	3350* *OARS WEEL			

SUBSTANCE / CAS#	ACGIH® TLVs® TWA ppm	TWA mg/m³	STEL/CEIL(C) ppm	STEL/CEIL(C) mg/m³	OSHA PELs TWA ppm	TWA mg/m³	STEL/CEIL(C) ppm	STEL/CEIL(C) mg/m³	NIOSH RELs TWA ppm	TWA mg/m³	STEL/CEIL(C) ppm	STEL/CEIL(C) mg/m³	DFG MAKs TWA ppm	TWA mg/m³	PEAK/CEIL(C) ppm	PEAK/CEIL(C) mg/m³	AIHA WEELs TWA ppm	TWA mg/m³	STEL/CEIL(C) ppm	STEL/CEIL(C) mg/m³	CARCINOGENICITY CATEGORY
1,1,1,3,3,3-Hexafluoropropane 690-39-1																	1000				
Hexafluoropropylene 116-15-4	0.1	0.6																			
Hexahydrophthalic acid diglycidyl ester 5493-45-8														Skin; Sh							MAK-3B
Hexahydrophthalic anhydride, all isomers 85-42-7; 13149-00-3; 14166-21-3			C 0.005 IFV											CAS: 85-42-7 Sa							
Hexamethylene bis(3-[3,5-di-*tert*-butyl-4-hydroxyphenyl]propionate) 35074-77-2														10 I C	II (2)						
1,6-Hexamethylene diisocyanate 822-06-0	0.005	0.034 BEI							0.005	0.035	C 0.02*	C 0.14* *10-min	0.005	0.035 Sah; D	I (1) C 0.01	C 0.07					
Hexamethylene glycol 629-11-8																		10			
Hexamethylenetetramine 100-97-0														Sh							

SUBSTANCE / CAS#	ACGIH® TLVs® TWA ppm	mg/m³	STEL/CEIL(C) ppm	mg/m³	OSHA PELs TWA ppm	mg/m³	STEL/CEIL(C) ppm	mg/m³	NIOSH RELs TWA ppm	mg/m³	STEL/CEIL(C) ppm	mg/m³	DFG MAKs TWA ppm	mg/m³	PEAK/CEIL(C) ppm	mg/m³	AIHA WEELs TWA ppm	mg/m³	STEL/CEIL(C) ppm	mg/m³	CARCINOGENICITY CATEGORY
Hexamethyl phos- phoramide 680-31-9			Skin						*See* Pocket Guide App. A					Skin; 2							IARC-2B TLV-A3 MAK-2 NIOSH-Ca NTP-R
n-Hexane (Hexane) 110-54-3	50	176	Skin; BEI		500	1800			50	180			50	180	II (8) C						EPA-II
Hexane, isomers, other than n-Hexane 75-83-2; 79-29-8; 96-14-0; 107-83-5	500	1760	1000	3500					100	350	C 510* *15-min	C 1800*	500	1800	II (2) including CAS: 96-37-7 D						
1,6-Hexanediamine 124-09-4	0.5	2.3															1				
1,6-Hexanediol diacrylate 13048-33-4														Sh			1	DSEN			
1-Hexene 592-41-6	50	172																			
sec-Hexyl acetate 108-84-9	50	295			50	300			50	300											
n-Hexyl alcohol 111-27-3																	40	eye irritation			
Hexylene glycol 107-41-5	25*		50* 10** *(V) **I (H)								C 25	C 125	10	49	I (2) D						

SUBSTANCE CAS#	ACGIH® TLVs® TWA ppm	ACGIH® TLVs® TWA mg/m³	ACGIH® TLVs® STEL/CEIL(C) ppm	ACGIH® TLVs® STEL/CEIL(C) mg/m³	OSHA PELs TWA ppm	OSHA PELs TWA mg/m³	OSHA PELs STEL/CEIL(C) ppm	OSHA PELs STEL/CEIL(C) mg/m³	NIOSH RELs TWA ppm	NIOSH RELs TWA mg/m³	NIOSH RELs STEL/CEIL(C) ppm	NIOSH RELs STEL/CEIL(C) mg/m³	DFG MAKs TWA ppm	DFG MAKs TWA mg/m³	DFG MAKs PEAK/CEIL(C) ppm	DFG MAKs PEAK/CEIL(C) mg/m³	AIHA WEELs TWA ppm	AIHA WEELs TWA mg/m³	AIHA WEELs STEL/CEIL(C) ppm	AIHA WEELs STEL/CEIL(C) mg/m³	CARCINOGENICITY CATEGORY
n-Hexyl mercaptan (n-Hexanethiol) 111-31-9											C 0.5* *15-min	C 2.7*									
HFE-7100 163702-07-6; 163702-08-7																	750				
Hycanthone mesylate 23255-93-8																					IARC-3
Hydralazine 86-54-4																					IARC-3
Hydrazine 302-01-2	0.01	0.013		Skin	1	1.3		Skin			C 0.03* *120-min See Pocket Guide App. A	C 0.04*		Skin; Sh							EPA-B2 NTP-R IARC-2B TLV-A3 MAK-2 NIOSH-Ca
Hydrazine hydrate and hydrazine salts 7803-57-8														Sh							
Hydrazine sulfate 10034-93-2																					NTP-R
Hydrazobenzene (1,2-Diphenylhydrazine) 122-66-7																					EPA-B2 MAK-2 NTP-R
Hydrazoic acid 7782-79-8													0.1	0.18	I (2)						

SUBSTANCE / CAS#	ACGIH® TLVs® TWA ppm	mg/m³	STEL/CEIL(C) ppm	mg/m³	OSHA PELs TWA ppm	mg/m³	STEL/CEIL(C) ppm	mg/m³	NIOSH RELs TWA ppm	mg/m³	STEL/CEIL(C) ppm	mg/m³	DFG MAKs TWA ppm	mg/m³	PEAK/CEIL(C) ppm	mg/m³	AIHA WEELs TWA ppm	mg/m³	STEL/CEIL(C) ppm	mg/m³	CARCINOGENICITY CATEGORY
Hydrochlorothiazide 58-93-5																					IARC-2B
Hydrogen 1333-74-0	*Documentation* withdrawn; *see* Appendix F in *TLVs® and BEIs®* book Simple asphyxiant (D, EX)																				
Hydrogenated ter-phenyls 61788-32-7	0.5	4.9 Nonirradiated							0.5	5											
Hydrogen bromide 10035-10-6			C 2	C 6.8	3	10					C 3	C 10	2	6.7	I (1) D						
Hydrogen chloride 7647-01-0			C 2	C 2.98			C 5	C 7			C 5	C 7	2	3	I (2) C						IARC-3 TLV-A4
Hydrogen cyanide 74-90-8			C 4.7* *as CN Skin	C 5** **Cyanide salts, as CN	10	11 Skin				4.7 Skin		5	1.9	2.1 Skin; C	II (2)						EPA-II* *and Cyanide salts
Hydrogen fluoride, as F 7664-39-3	0.5	0.41 Skin; BEI	C 2	C 1.64	3				3	2.5	C 6* *15-min	C 5*	1	0.83 C	I (2)						
Hydrogen peroxide 7722-84-1	1	1.4			1	1.4			1	1.4			0.5	0.71 C	I (1)						IARC-3 MAK-4 TLV-A3
Hydrogen selenide 7783-07-5	0.05	0.16			0.05* *as Se	0.2*			0.05	0.2			0.006	0.02 C	II (8)						MAK-3B

SUBSTANCE CAS#	ACGIH® TLVs® TWA ppm	TWA mg/m³	STEL/CEIL(C) ppm	mg/m³	OSHA PELs TWA ppm	TWA mg/m³	STEL/CEIL(C) ppm	mg/m³	NIOSH RELs TWA ppm	TWA mg/m³	STEL/CEIL(C) ppm	mg/m³	DFG MAKs TWA ppm	TWA mg/m³	PEAK/CEIL(C) ppm	mg/m³	AIHA WEELs TWA ppm	TWA mg/m³	STEL/CEIL(C) ppm	mg/m³	CARCINOGENICITY CATEGORY
Hydrogen sulfide 7783-06-4	1	1.4	5	7			C 20; 50*				C 10*	C 15*	5	7.1	I (2)						EPA-I
					*10-min peak; once per 8-hr shift				*10-min				C								
Hydroquinone (Dihydroxybenzene) 123-31-9		1			2						C 2*										IARC-3 MAK-2 TLV-A3
		DSEN							*15-min				Skin; Sh; 3A								
1-Hydroxyanthra- quinone 129-43-1																					IARC-2B
4-Hydroxyazobenzene 1689-82-3																					IARC-3
Hydroxybenzoic acid 99-96-7																		5			
Hydroxycitronellal 107-75-5														Sh							
2-Hydroxyethyl acrylate 818-61-1														Sh							
N,N′,N″-tris(β-Hydroxy-ethyl)-hexahydro-1,3,5-triazine 4719-04-4														Sh							
Hydroxylamine and its salts 7803-49-8														Sh							

SUBSTANCE / CAS#	ACGIH® TLVs® TWA ppm	TWA mg/m³	STEL/CEIL(C) ppm	STEL/CEIL(C) mg/m³	OSHA PELs TWA ppm	TWA mg/m³	STEL/CEIL(C) ppm	STEL/CEIL(C) mg/m³	NIOSH RELs TWA ppm	TWA mg/m³	STEL/CEIL(C) ppm	STEL/CEIL(C) mg/m³	DFG MAKs TWA ppm	TWA mg/m³	PEAK/CEIL(C) ppm	PEAK/CEIL(C) mg/m³	AIHA WEELs TWA ppm	TWA mg/m³	STEL/CEIL(C) ppm	STEL/CEIL(C) mg/m³	CARCINOGENICITY CATEGORY
4-(4-Hydroxy-4-methylpentyl)-3-cyclohexene-1-carboxaldehyde (Lyral) 31906-04-4													Sh								
Hydroxypropyl acrylate, all isomers (Acrylic acid hydroxypropyl ester) 25584-83-2													Sh								
2-Hydroxypropyl acrylate 999-61-1	0.5	2.8							0.5	3			Sh								
		Skin; DSEN								Skin											
2-Hydroxypropyl methacrylate (Methacrylic acid 2-hydroxypropyl ester) 923-26-2													Sh								
8-Hydroxyquinoline 148-24-3																					IARC-3
Hydroxysenkirkine 26782-43-4																					IARC-3
Hydroxyurea 127-07-1																					IARC-3
Hypochlorite salts																					IARC-3
Indene 95-13-6	5	24							10	45											

SUBSTANCE / CAS#	ACGIH® TLVs® TWA ppm	mg/m³	STEL/CEIL(C) ppm	mg/m³	OSHA PELs TWA ppm	mg/m³	STEL/CEIL(C) ppm	mg/m³	NIOSH RELs TWA ppm	mg/m³	STEL/CEIL(C) ppm	mg/m³	DFG MAKs TWA ppm	mg/m³	PEAK/CEIL(C) ppm	mg/m³	AIHA WEELs TWA ppm	mg/m³	STEL/CEIL(C) ppm	mg/m³	CARCINOGENICITY CATEGORY
Indeno[1,2,3,cd]pyrene 193-39-5														Skin							EPA-B2 IARC-2B MAK-2 NTP-R
Indium [7440-74-6] and compounds, as In		0.1								0.1											
Indium phosphide 22398-80-7																					IARC-2A MAK-2
Iodides		0.01 **IFV**																			TLV-A4
Iodine 7553-56-2	0.01*	0.1* ***IFV**	0.1 (V)	1			C 0.1	C 1			C 0.1	C 1		Skin							TLV-A4
Iodoform 75-47-8	(0.6) NIC-0.2*	(10) NIC-3.3* ***IFV**							0.6	10											
3-Iodo-2-propynyl butylcarbamate 55406-53-6													0.005	0.058* *can also occur as vapor and aerosol	I (2) Sh; C						
Iron-dextran complex 9004-66-4																					IARC-2B NTP-R
Iron-dextrin complex 9004-51-7																					IARC-3

SUBSTANCE / CAS#	ACGIH® TLVs® TWA ppm	ACGIH® TLVs® TWA mg/m³	ACGIH® TLVs® STEL/CEIL(C) ppm	ACGIH® TLVs® STEL/CEIL(C) mg/m³	OSHA PELs TWA ppm	OSHA PELs TWA mg/m³	OSHA PELs STEL/CEIL(C) ppm	OSHA PELs STEL/CEIL(C) mg/m³	NIOSH RELs TWA ppm	NIOSH RELs TWA mg/m³	NIOSH RELs STEL/CEIL(C) ppm	NIOSH RELs STEL/CEIL(C) mg/m³	DFG MAKs TWA ppm	DFG MAKs TWA mg/m³	DFG MAKs PEAK/CEIL(C) ppm	DFG MAKs PEAK/CEIL(C) mg/m³	AIHA WEELs TWA ppm	AIHA WEELs TWA mg/m³	AIHA WEELs STEL/CEIL(C) ppm	AIHA WEELs STEL/CEIL(C) mg/m³	CARCINOGENICITY CATEGORY
Iron oxide (FeO) 1345-25-1													with the exception of Iron oxides which are not biologically available								MAK-3B
Iron oxide (Fe₂O₃) 1309-37-1		5 R				10* *Fume				5* *Dust and fume, as Fe			with the exception of Iron oxides which are not biologically available								IARC-3 MAK-3B TLV-A4
Iron pentacarbonyl 13463-40-6	0.1	0.23	0.2	0.45					0.1*	0.23* *as Fe	0.2*	0.45*	0.1	0.81 Skin; D	I (2)						
Iron salts, soluble, as Fe		1								1											
Iron sorbitol-citric acid complex 1338-16-5																					IARC-3
Isatidine 15503-86-3																					IARC-3
Isoamyl alcohol 123-51-3	100	361	125	452	100*	360* *primary and secondary			100*	360* *primary and secondary	125*	450*	20	73 C	I (4)						
Isobutanol (Isobutyl alcohol) 78-83-1	50	152			100	300			50	150			100	310 C	I (1)						
Isobutyl acetate 110-19-0	Adopted TLV® and *Documentation* withdrawn; *see* Butyl acetates, all isomers				150	700			150	700			100	480 C	I (2)						

SUBSTANCE / CAS#	ACGIH® TLVs® TWA ppm	TWA mg/m³	STEL/CEIL(C) ppm	STEL/CEIL(C) mg/m³	OSHA PELs TWA ppm	TWA mg/m³	STEL/CEIL(C) ppm	STEL/CEIL(C) mg/m³	NIOSH RELs TWA ppm	TWA mg/m³	STEL/CEIL(C) ppm	STEL/CEIL(C) mg/m³	DFG MAKs TWA ppm	TWA mg/m³	PEAK/CEIL(C) ppm	PEAK/CEIL(C) mg/m³	AIHA WEELs TWA ppm	TWA mg/m³	STEL/CEIL(C) ppm	STEL/CEIL(C) mg/m³	CARCINOGENICITY CATEGORY
Isobutylamine 78-81-9													2	6.1	I (2) C 5 C 15 D						
Isobutyl nitrite 542-56-3			C 1 **IFV**	C 4.2 **IFV** BEI$_M$																	TLV-A3
Isobutyraldehyde 78-84-2																	25				
Isobutyronitrile 78-82-0									8	22											
Isocyanuric acid 108-80-5																	10* 5 **R** *Total dust				
Isoeugenol and its isomers 97-54-1; 5912-86-7; 5932-68-3														Sh							
Isonicotinic acid hydrazine (Isoniazid) 54-85-3																					IARC-3
Isooctyl acrylate (2-Propenoic acid, isooctyl ester) 29590-42-9																	5				
Isooctyl alcohol 26952-21-6	50	266 Skin							50	270 Skin											

SUBSTANCE / CAS#	ACGIH® TLVs® TWA ppm	mg/m³	STEL/CEIL(C) ppm	mg/m³	OSHA PELs TWA ppm	mg/m³	STEL/CEIL(C) ppm	mg/m³	NIOSH RELs TWA ppm	mg/m³	STEL/CEIL(C) ppm	mg/m³	DFG MAKs TWA ppm	mg/m³	PEAK/CEIL(C) ppm	mg/m³	AIHA WEELs TWA ppm	mg/m³	STEL/CEIL(C) ppm	mg/m³	CARCINOGENICITY CATEGORY
Isopentyl acetate (Isoamyl acetate) 123-92-2	50	266	100	532	100	525			100	525			50	270	I (1) D						
Isophorone 78-59-1			C 5	C 28	25	140			4	23			2	11	I (2) C						EPA-C MAK-3B TLV-A3
Isophorone diisocyanate 4098-71-9	0.005	0.045							0.005	0.045	0.02	0.18 Skin	0.005	0.046	I (1) C 0.01 C 0.092 Sah; D						
Isophosphamide 3778-73-2																					IARC-3
Isoprene 78-79-5													3	8.5	II (8) C; 5		2				IARC-2B MAK-5 NTP-R
Isopropenyl acetate 108-22-5													10	46	I (2) D						
2-Isopropoxyethanol (Ethylene glycol isopropyl ether) 109-59-1	25	106 Skin											5	22	II (8) Skin; C						
Isopropyl acetate 108-21-4	(100)	(418)	(200)	(836)	250	950							100	420	I (2) C						
	NIC-withdraw adopted TLVs® and Documentation; see Propyl acetate isomers																				
Isopropylamine 75-31-0	5	12	10	24	5	12							5	12	I (2) C 10 C 25 C						

SUBSTANCE CAS#	ACGIH® TLVs® TWA ppm	mg/m³	STEL/CEIL(C) ppm	mg/m³	OSHA PELs TWA ppm	mg/m³	STEL/CEIL(C) ppm	mg/m³	NIOSH RELs TWA ppm	mg/m³	STEL/CEIL(C) ppm	mg/m³	DFG MAKs TWA ppm	mg/m³	PEAK/CEIL(C) ppm	mg/m³	AIHA WEELs TWA ppm	mg/m³	STEL/CEIL(C) ppm	mg/m³	CARCINOGENICITY CATEGORY
N-Isopropylaniline 768-52-5	2	11							2	10											
	Skin; BEI$_M$								Skin												
Isopropyl ether 108-20-3	250	1040	310	1300	500	2100			500	2100			200	850	I (2)						
														C							
Isopropyl glycidyl ether (IGE) 4016-14-2	50	238	75	356	50	240					C 50*	C 240*									MAK-3B
											*15-min										
Isopropyl methyl phosphonic acid (IMPA) 1832-54-8																					EPA-D
Isopropyl oil (residue of Isopropyl alcohol production)																					IARC-3 MAK-3B
4-Isopropylphenyl isocyanate 31027-31-3														Sh							
N-Isopropyl-N'-phenyl-p-phenylenediamine 101-72-4													2 I		II (2)						
												Sh; C									
Isosafrole 120-58-1																					IARC-3
Isoxaben 82558-50-7																					EPA-C

SUBSTANCE / CAS#	ACGIH® TLVs® TWA ppm	mg/m³	STEL/CEIL(C) ppm	mg/m³	OSHA PELs TWA ppm	mg/m³	STEL/CEIL(C) ppm	mg/m³	NIOSH RELs TWA ppm	mg/m³	STEL/CEIL(C) ppm	mg/m³	DFG MAKs TWA ppm	mg/m³	PEAK/CEIL(C) ppm	mg/m³	AIHA WEELs TWA ppm	mg/m³	STEL/CEIL(C) ppm	mg/m³	CARCINOGENICITY CATEGORY
Jacobine 6870-67-3																					IARC-3
Kaolin 1332-58-7		2 **R** E			15*; 5** *Total dust	**Respirable fraction			10*; 5** *Total dust	**Respirable fraction											MAK-3B* TLV-A4 *Quartz content must be considered separately
Kempferol 520-18-3																					IARC-3
Kerosene/Jet fuels as total hydrocarbon vapor 8008-20-6; 64742-81-0		200 Skin; P								100 Kerosene only											IARC-3* TLV-A3 *Jet fuels only
Ketene 463-51-4	0.5	0.86	1.5	2.6	0.5	0.9			0.5	0.9	1.5	3									
Kevlar 49 (p-Aramid fibrils) 24938-64-5																					IARC-3
Kojic acid 501-30-4																					IARC-3
Lasiocarpine 303-34-4																					IARC-2B
Lauric acid 143-07-7															2 **I*** *can also occur as vapor and aerosol D	I (2)					MAK-3A

SUBSTANCE CAS#	ACGIH® TLVs® TWA ppm	ACGIH® TLVs® TWA mg/m³	ACGIH® TLVs® STEL/CEIL(C) ppm	ACGIH® TLVs® STEL/CEIL(C) mg/m³	OSHA PELs TWA ppm	OSHA PELs TWA mg/m³	OSHA PELs STEL/CEIL(C) ppm	OSHA PELs STEL/CEIL(C) mg/m³	NIOSH RELs TWA ppm	NIOSH RELs TWA mg/m³	NIOSH RELs STEL/CEIL(C) ppm	NIOSH RELs STEL/CEIL(C) mg/m³	DFG MAKs TWA ppm	DFG MAKs TWA mg/m³	DFG MAKs PEAK/CEIL(C) ppm	DFG MAKs PEAK/CEIL(C) mg/m³	AIHA WEELs TWA ppm	AIHA WEELs TWA mg/m³	AIHA WEELs STEL/CEIL(C) ppm	AIHA WEELs STEL/CEIL(C) mg/m³	CARCINOGENICITY CATEGORY
Lauroyl peroxide 105-74-8																					IARC-3
Lead [7439-92-1] **and inorganic compounds, as Pb**	0.05				0.05 including organic Lead soaps See 29 CFR 1910.1025				0.05* *8-hr TWA excluding Lead arsenate See Pocket Guide App. C				except Lead arsenate and Lead chromate; as Inhalable fraction 3A								EPA-B2 NTP-R IARC-2A*; 2B TLV-A3 MAK-2 *inorganic compounds
		BEI																			
Lead, organic compounds					for organic Lead soaps, see Lead and inorganic compounds, as Pb																IARC-3
Lead arsenate 3687-31-8	TLV® withdrawn due to insufficient data				0.01* *as As See 29 CFR 1910.1018				See Arsenic and inorganic compounds				as As Skin; 3A								EPA-B2 NTP-R IARC-1; 2A OSHA-Ca MAK-1
Lead chromate 7758-97-6	(0.05*) (0.012**) NIC-0.0002 I *as Pb **as Cr NIC-A1 NIC-DSEN; RSEN		NIC-0.0005 I						See Chromium(VI) inorganic compounds, insoluble				See Chromium(VI) inorganic compounds, insoluble								EPA*-A**; (TLV-A2) D***; CBD**; K** NTP-K* *as Cr(VI) **inhalation ***oral
Lead chromate oxide 18454-12-1													See Chromium(VI) inorganic compounds, insoluble								
Lead phosphate 7446-27-7	See Lead and inorganic compounds, as Pb				0.05 See 29 CFR 1910.1025				See Lead and inorganic compounds												EPA-B2 IARC-2A NTP-R TLV-A3

SUBSTANCE / CAS#	ACGIH® TLVs® TWA ppm	TWA mg/m³	STEL/CEIL(C) ppm	STEL/CEIL(C) mg/m³	OSHA PELs TWA ppm	TWA mg/m³	STEL/CEIL(C) ppm	STEL/CEIL(C) mg/m³	NIOSH RELs TWA ppm	TWA mg/m³	STEL/CEIL(C) ppm	STEL/CEIL(C) mg/m³	DFG MAKs TWA ppm	TWA mg/m³	PEAK/CEIL(C) ppm	PEAK/CEIL(C) mg/m³	AIHA WEELs TWA ppm	TWA mg/m³	STEL/CEIL(C) ppm	STEL/CEIL(C) mg/m³	CARCINOGENICITY CATEGORY
Levofuraltadone (5-[Morpholinomethyl]-3-[(5-nitrofurfurylidene)amino]-2-oxazolidinone) 3795-88-8																					IARC-2B
Light Green SF 5141-20-8																					IARC-3
D-Limonene 5989-27-5													5	28	II (4)						IARC-3
													Skin; Sh; C								
DL-Limonene 138-86-3													and similar mixtures Sh				30				
L-Limonene (β-Limonene) 5989-54-8													Sh								
Lindane (γ-Hexachlorocyclohexane) 58-89-9	0.5				0.5				0.5				0.1 I		II (8)						MAK-4 TLV-A3 NTP-R*
	Skin				Skin				Skin				Skin; C								*and other HCH isomers
Linuron 330-55-2																					EPA-C

SUBSTANCE / CAS#	ACGIH® TLVs® TWA ppm	mg/m³	STEL/CEIL(C) ppm	mg/m³	OSHA PELs TWA ppm	mg/m³	STEL/CEIL(C) ppm	mg/m³	NIOSH RELs TWA ppm	mg/m³	STEL/CEIL(C) ppm	mg/m³	DFG MAKs TWA ppm	mg/m³	PEAK/CEIL(C) ppm	mg/m³	AIHA WEELs TWA ppm	mg/m³	STEL/CEIL(C) ppm	mg/m³	CARCINOGENICITY CATEGORY
Lithium [7439-93-2] compounds, inorganic, as Li, except of Li and highly irritating lithium compounds (as lithium amide, hydride, hydroxide, nitride, oxide, tetrahydroaluminate, tetrahydroborate)														0.2 I		I (1) C					
Lithium hydride 7580-67-8			C 0.05 I			0.025				0.025											
Lithium hydroxide 1310-65-2																				C 1	
Lithium oxide 12057-24-8																				C 1	
L.P.G. (Liquefied petroleum gas) 68476-85-7	*See* Appendix F in *TLVs®* and *BEIs®* book (D, EX)				1000	1800			1000	1800											
Luteoskyrin 21884-44-6																					IARC-3
Magenta 632-99-5																					IARC-1*; 2B *production

SUBSTANCE / CAS#	ACGIH® TLVs® TWA ppm	TWA mg/m³	STEL/CEIL(C) ppm	STEL/CEIL(C) mg/m³	OSHA PELs TWA ppm	TWA mg/m³	STEL/CEIL(C) ppm	STEL/CEIL(C) mg/m³	NIOSH RELs TWA ppm	TWA mg/m³	STEL/CEIL(C) ppm	STEL/CEIL(C) mg/m³	DFG MAKs TWA ppm	TWA mg/m³	PEAK/CEIL(C) ppm	PEAK/CEIL(C) mg/m³	AIHA WEELs TWA ppm	TWA mg/m³	STEL/CEIL(C) ppm	STEL/CEIL(C) mg/m³	CARCINOGENICITY CATEGORY
Magnesite 546-93-0	TLV® withdrawn due to insufficient data				15*; 5** *Total dust **Respirable fraction				10*; 5** *Total dust **Respirable fraction												
Magnesium oxide 1309-48-4		10 I			15* Fume *Total particulate									4 I 1.5 R C							TLV-A4
Magnesium oxide sulfate, fibrous dust 12286-12-3																					MAK-3B
Malathion 121-75-5	1 IFV Skin; BEI_A				15* *Total dust Skin				10 Skin				15 I D		II (4)						IARC-3 TLV-A4
Maleic anhydride 108-31-6	0.0025* *IFV DSEN; RSEN	0.01*			0.25	1			0.25	1			0.1 Sah; C	0.41	I (1) C 0.2	C 0.81					TLV-A4
Maleic hydrazide 123-33-1																					IARC-3
Malonaldehyde 542-78-9									*See* Pocket Guide Apps. A and C												IARC-3 NIOSH-Ca
Malononitrile 109-77-3									3	8											
Mancozeb 8018-01-7																		1 DSEN			

SUBSTANCE / CAS#	ACGIH® TLVs® TWA ppm	mg/m³	STEL/CEIL(C) ppm	mg/m³	OSHA PELs TWA ppm	mg/m³	STEL/CEIL(C) ppm	mg/m³	NIOSH RELs TWA ppm	mg/m³	STEL/CEIL(C) ppm	mg/m³	DFG MAKs TWA ppm	mg/m³	PEAK/CEIL(C) ppm	mg/m³	AIHA WEELs TWA ppm	mg/m³	STEL/CEIL(C) ppm	mg/m³	CARCINOGENICITY CATEGORY
Manganese [7439-96-5] and inorganic compounds, as Mn		0.02 **R** 0.1 **I**						C 5	1			3		0.2 **I** 0.02 **R**	II (8) II (1)* *Permanganates only C						EPA-D TLV-A4
Manganese, fume, as Mn 7439-96-5		0.02 **R** 0.1 **I**						C 5	1			3		0.2 **I** 0.02 **R** C	II (8)						EPA-D
Manganese cyclopenta-dienyl tricarbonyl, as Mn 12079-65-1		0.1 Skin						C 5	0.1		Skin			0.2 **I** 0.02 **R** C	II (8)						
Manganous ethylenebis-(dithiocarbamate) (Maneb) 12427-38-2														Sh							IARC-3
Mannomustine dihydrochloride 551-74-6																					IARC-3
Mate, absolute (Tea oil) 68916-96-1																					IARC-3
Medphalan 13045-94-8																					IARC-3
Medroxyprogesterone acetate 71-58-9																					IARC-2B
Melamine 108-78-1																		3* *OARS WEEL			IARC-3

SUBSTANCE / CAS#	ACGIH® TLVs® TWA ppm	TWA mg/m³	STEL/CEIL(C) ppm	STEL/CEIL(C) mg/m³	OSHA PELs TWA ppm	TWA mg/m³	STEL/CEIL(C) ppm	STEL/CEIL(C) mg/m³	NIOSH RELs TWA ppm	TWA mg/m³	STEL/CEIL(C) ppm	STEL/CEIL(C) mg/m³	DFG MAKs TWA ppm	TWA mg/m³	PEAK/CEIL(C) ppm	PEAK/CEIL(C) mg/m³	AIHA WEELs TWA ppm	TWA mg/m³	STEL/CEIL(C) ppm	STEL/CEIL(C) mg/m³	CARCINOGENICITY CATEGORY
Melphalan 148-82-3																					IARC-1 NTP-K
Menthol, raw, natural, synthetic 89-78-1; 1490-04-6; 2216-51-5; 15356-70-4																	1*	6.4*	3*	19.2* *OARS WEEL	
Merbromin 129-16-8														Sh							
2-Mercaptobenzo-thiazole 149-30-4														4 I 1.5 R Sh; C				5 Skin; DSEN			MAK-3B
Mercaptoethanol 60-24-2																	0.2	Skin			
6-Mercaptopurine 50-44-2																					IARC-3
Mercuric chloride 7487-94-7																					EPA-C
Mercury, alkyl compounds, as Hg	0.01		0.03	Skin	0.01		C 0.04		0.01		0.03	Skin		Skin; Sh							MAK-3B
Mercury, aryl compounds, as Hg 7439-97-6	0.1			Skin	0.1 See OSHA standard interpretation memo						C 0.1	Skin		Skin; Sh							MAK-3B

SUBSTANCE CAS#	ACGIH® TLVs® TWA ppm	mg/m³	STEL/CEIL(C) ppm	mg/m³	OSHA PELs TWA ppm	mg/m³	STEL/CEIL(C) ppm	mg/m³	NIOSH RELs TWA ppm	mg/m³	STEL/CEIL(C) ppm	mg/m³	DFG MAKs TWA ppm	mg/m³	PEAK/CEIL(C) ppm	mg/m³	AIHA WEELs TWA ppm	mg/m³	STEL/CEIL(C) ppm	mg/m³	CARCINOGENICITY CATEGORY
Mercury, elemental and inorganic compounds, as Hg 7439-97-6		0.025				0.1 *See* OSHA standard interpretation memo					0.05* *Vapor	C 0.1			0.02 I	II (8)						EPA-D IARC-3 MAK-3B TLV-A4
	Skin; BEI								Skin				Skin; Sh; D									
Merphalan 531-76-0																						IARC-2B
Mesityl oxide 141-79-7	15	60	25	100	25	100			10	40			2	8.1	I (2)							
													Skin; D									
Metabisulfites 23134-05-6																						IARC-3
Metal-working fluids (capable of yielding nitrosamines)																						MAK-3B
Methacrylic acid 79-41-4	20	70							20	70			50	180	I (2)							
									Skin				C									
Methane 74-82-8	(*See* Appendix F in *TLVs*® and *BEIs*® book) NIC-D, EX																					
Methanol (Methyl alcohol) 67-56-1	200	262	250	328	200	260			200	260	250	325	200	270	II (4)							
	Skin; BEI								Skin				Skin; C									
Methenamin 3-chlor-allylchloride 4080-31-3													2 I releases Formaldehyde Sh; B		II (2)							

SUBSTANCE / CAS#	ACGIH® TLVs® TWA ppm	mg/m³	STEL/CEIL(C) ppm	mg/m³	OSHA PELs TWA ppm	mg/m³	STEL/CEIL(C) ppm	mg/m³	NIOSH RELs TWA ppm	mg/m³	STEL/CEIL(C) ppm	mg/m³	DFG MAKs TWA ppm	mg/m³	PEAK/CEIL(C) ppm	mg/m³	AIHA WEELs TWA ppm	mg/m³	STEL/CEIL(C) ppm	mg/m³	CARCINOGENICITY CATEGORY
Methidathion 950-37-8																					EPA-C
Methimazole 60-56-0																					IARC-3
Methomyl 16752-77-5		0.2 **IFV** Skin; BEI$_A$								2.5											TLV-A4
Methotrexate 59-05-2																					IARC-3
Methoxyacetic acid 625-45-6													1	3.7 Skin; B	II (2)						
Methoxychlor 72-43-5		10				15* *Total dust				*See* Pocket Guide App. A				1 **I** Skin; B	II (8)						EPA-D IARC-3 NIOSH-Ca TLV-A4
2-Methoxyethanol (EGME) 109-86-4	0.1	0.3 Skin; BEI			25	80 Skin			0.1	0.3 Skin			1*	3.2 *sum of the concentrations of EGME and its acetate in air Skin; B	II (8)			Skin; B			
2-Methoxyethyl acetate (EGMEA) 110-49-6	0.1	0.5 Skin; BEI			25	120 Skin			0.1	0.5 Skin			1*	4.9 *sum of the concentrations of EGME and its acetate in air Skin; B	II (8)			Skin; B			
Methoxyflurane 76-38-0											C 2* *60-min; for exposure to waste anesthetic gases	C 13.5*									

SUBSTANCE CAS#	ACGIH® TLVs® TWA ppm	mg/m³	STEL/CEIL(C) ppm	mg/m³	OSHA PELs TWA ppm	mg/m³	STEL/CEIL(C) ppm	mg/m³	NIOSH RELs TWA ppm	mg/m³	STEL/CEIL(C) ppm	mg/m³	DFG MAKs TWA ppm	mg/m³	PEAK/CEIL(C) ppm	mg/m³	AIHA WEELs TWA ppm	mg/m³	STEL/CEIL(C) ppm	mg/m³	CARCINOGENICITY CATEGORY
(2-Methoxymethylethoxy)-propanol (DPGME) 34590-94-8	100	606	150	909	100	600			100	600	150	900	50	310	I (1)						
		Skin				Skin				Skin				D							
4-Methoxyphenol 150-76-5		5								5											
1-Methoxy-2-propanol (Propylene glycol monomethyl ether; PGME) 107-98-2	50	184	100	369					100	360	150	540	100	370	I (2)						TLV-A4
															C						
2-Methoxy-1-propanol (Propylene glycol 2-methyl ether) 1589-47-5													5	19	II (8)						
														Skin; B							
1-Methoxypropyl-2-acetate (Propylene glycol monomethyl ether acetate) 108-65-6													50	270	I (1)		50				
														C							
2-Methoxypropyl-1-acetate (Propylene glycol 2-methyl ether-1-acetate) 70657-70-4													5	28	II (8)						
														Skin; B							
3-Methoxypropylamine 5332-73-0																	5	15			
5-Methoxypsoralen 484-20-8																					IARC-2A
8-Methoxypsoralen (Methoxsalen) plus ultraviolet A radiation 298-81-7																					IARC-1 NTP-K

SUBSTANCE / CAS#	ACGIH® TLVs® TWA ppm	mg/m³	STEL/CEIL(C) ppm	mg/m³	OSHA PELs TWA ppm	mg/m³	STEL/CEIL(C) ppm	mg/m³	NIOSH RELs TWA ppm	mg/m³	STEL/CEIL(C) ppm	mg/m³	DFG MAKs TWA ppm	mg/m³	PEAK/CEIL(C) ppm	mg/m³	AIHA WEELs TWA ppm	mg/m³	STEL/CEIL(C) ppm	mg/m³	CARCINOGENICITY CATEGORY
Methyl acetate 79-20-9	200	606	250	757	200	610			200	610	250	760	100	310	I (4) C						
Methylacetylene (Propyne) 74-99-7	1000* *(EX)	1640*			1000	1650			1000	1650											
Methylacetylene-propadiene mixture (MAPP)	1000* *(EX)	1640*	1250*	2050*	1000	1800			1000	1800	1250	2250									
Methyl acrylate (Acrylic acid, methyl ester) 96-33-3	2	7			10	35			10	35			2	7.1	I (2)						EPA-D IARC-3 TLV-A4
		Skin; DSEN				Skin				Skin				Skin; Sh; C							
Methylacrylonitrile 126-98-7	1	2.7							1	3											TLV-A4
		Skin								Skin											
Methylal (Dimethoxymethane) 109-87-5	1000	3110			1000	3100			1000	3100			1000	3200	II (2) C						
Methylamine 74-89-5	5	6.4	15	19	10	12			10	12			10	13	I (1) C 10 C 13 D						
Methyl n-amyl ketone (2-Heptanone) 110-43-0	50	233			100	465			100	465											
5-Methylangelicin plus ultraviolet A radiation 73459-03-7																					IARC-3

SUBSTANCE CAS#	ACGIH® TLVs® TWA ppm	TWA mg/m³	STEL/CEIL(C) ppm	STEL/CEIL(C) mg/m³	OSHA PELs TWA ppm	TWA mg/m³	STEL/CEIL(C) ppm	STEL/CEIL(C) mg/m³	NIOSH RELs TWA ppm	TWA mg/m³	STEL/CEIL(C) ppm	STEL/CEIL(C) mg/m³	DFG MAKs TWA ppm	TWA mg/m³	PEAK/CEIL(C) ppm	PEAK/CEIL(C) mg/m³	AIHA WEELs TWA ppm	TWA mg/m³	STEL/CEIL(C) ppm	STEL/CEIL(C) mg/m³	CARCINOGENICITY CATEGORY
N-Methylaniline (Monomethyl aniline) 100-61-8	0.5	2.2			2	9			0.5	2			0.5	2.2	II (2)						MAK-3B
	Skin; BEI_M				Skin				Skin				Skin; D								
Methylarsenic compounds																					MAK-1
													Skin; 3A								
Methylarsonic acid 124-58-3																					IARC-2B
tris(2-Methyl-1-aziridinyl)phosphine oxide 57-39-6																					IARC-3
Methylazoxymethanol acetate 592-62-1																					IARC-2B
Methyl bromide 74-83-9	1	3.9					C 20	C 80					1	3.9	I (2)						EPA-D TLV-A4 IARC-3 MAK-3B NIOSH-Ca
	Skin				Skin				See Pocket Guide App. A				C								
2-Methylbutyl acetate 624-41-9	50	266	100	532									50	270	I (1)						
													C								
Methyl tert-butyl ether (MTBE) 1634-04-4	50	180											50	180	I (1.5)						IARC-3 MAK-3B TLV-A3
													C								
Methyl n-butyl ketone (2-Hexanone) 591-78-6	5	20	10	40	100	410			1	4			5	21	II (8)						EPA-I
	Skin; BEI												Skin								

SUBSTANCE / CAS#	ACGIH® TLVs® TWA ppm	mg/m³	STEL/CEIL(C) ppm	mg/m³	OSHA PELs TWA ppm	mg/m³	STEL/CEIL(C) ppm	mg/m³	NIOSH RELs TWA ppm	mg/m³	STEL/CEIL(C) ppm	mg/m³	DFG MAKs TWA ppm	mg/m³	PEAK/CEIL(C) ppm	mg/m³	AIHA WEELs TWA ppm	mg/m³	STEL/CEIL(C) ppm	mg/m³	CARCINOGENICITY CATEGORY
Methyl carbamate 598-55-0																					IARC-3
Methyl chloride 74-87-3	50	103	100	207	100		C 200; 300* *5-min peak in any 3 hrs		*See* Pocket Guide App. A				50	100	II (2) Skin; B						EPA-D; CBD IARC-3 MAK-3B NIOSH-Ca TLV-A4
Methyl chloroform (1,1,1-Trichloroethane) 71-55-6	350	1910 BEI	450	2460	350	1900					C 350* C 1900* *15-min *See* Pocket Guide App. C		200	1100	II (1) Skin; C						EPA-II IARC-3 TLV-A4
Methyl chloroformate (Chloroformic acid methyl ester) 79-22-1													0.2	0.78	I (2) C						
1-Methylchrysene 3351-28-8																					IARC-3
2-Methylchrysene 3351-32-4																					IARC-3
3-Methylchrysene 3351-31-3																					IARC-3
4-Methylchrysene 3351-30-2																					IARC-3
5-Methylchrysene 3697-24-3																					IARC-2B NTP-R

SUBSTANCE / CAS#	ACGIH® TLVs® TWA ppm	mg/m³	STEL/CEIL(C) ppm	mg/m³	OSHA PELs TWA ppm	mg/m³	STEL/CEIL(C) ppm	mg/m³	NIOSH RELs TWA ppm	mg/m³	STEL/CEIL(C) ppm	mg/m³	DFG MAKs TWA ppm	mg/m³	PEAK/CEIL(C) ppm	mg/m³	AIHA WEELs TWA ppm	mg/m³	STEL/CEIL(C) ppm	mg/m³	CARCINOGENICITY CATEGORY
6-Methylchrysene 1705-85-7																					IARC-3
Methyl-2-cyano-acrylate 137-05-3	(0.2)	(1.0)	NIC-withdraw adopted TLV® and *Documentation; see* Cyanoacrylates, Ethyl and Methyl						2	8	4	16	2	9.2	I (1) D						
Methylcyclohexane 108-87-2	400	1610			500	2000			400	1600			200	810	II (2) D						
Methylcyclohexanol 25639-42-3	50	234			100	470			50	235											
o-Methylcyclohexanone 583-60-8	50	229	75	344	100	460			50	230	75	345									
	Skin				Skin				Skin												
2-Methylcyclopentadienyl manganese tricarbonyl, as Mn 12108-13-3		0.2								0.2											
	Skin								Skin												
Methyl demeton (Demeton-methyl) 8022-00-2		0.05 IFV								0.5			0.5	4.8	II (2)						
	Skin; BEIₐ								Skin				Skin								
N-Methyl-N,4-dinitroso-aniline 99-80-9																					IARC-3
Methylene bisphenyl iso-cyanate (MDI; Diphenylmethane-4,4'-diisocyanate) 101-68-8	0.005	0.051					C 0.02	C 0.2	0.005	0.05	C 0.02*	C 0.2*		0.05 I	I (1)	C 0.1					EPA-CBD; D IARC-3 MAK-4
											*10-min		Skin; Sah; C								

SUBSTANCE / CAS#	ACGIH® TLVs® TWA ppm	TWA mg/m³	STEL/CEIL(C) ppm	STEL/CEIL(C) mg/m³	OSHA PELs TWA ppm	TWA mg/m³	STEL/CEIL(C) ppm	STEL/CEIL(C) mg/m³	NIOSH RELs TWA ppm	TWA mg/m³	STEL/CEIL(C) ppm	STEL/CEIL(C) mg/m³	DFG MAKs TWA ppm	TWA mg/m³	PEAK/CEIL(C) ppm	PEAK/CEIL(C) mg/m³	AIHA WEELs TWA ppm	TWA mg/m³	STEL/CEIL(C) ppm	STEL/CEIL(C) mg/m³	CARCINOGENICITY CATEGORY
4,4'-Methylene bis(2-chloroaniline) (MBOCA) 101-14-4	0.01	0.11							0.003 Skin *See* Pocket Guide App. A				Skin								IARC-1 TLV-A2 MAK-2 NIOSH-Ca NTP-R
		Skin; BEI																			
Methylene bis(4-cyclo-hexylisocyanate) 5124-30-1	0.005	0.054								C 0.01	C 0.11			Sh							
4,4'-Methylenedianiline (4,4'-Diaminodiphenyl-methane) 101-77-9	0.1	0.81			0.01		0.1						Skin; Sh								IARC-2B NTP-R* MAK-2 OSHA-Ca NIOSH-Ca TLV-A3 *and its salts; CAS: 13552-44-8
		Skin			*See* 29 CFR 1910.1050				*See* Pocket Guide App. A												
4,4'-Methylene bis(N,N'-dimethyl)aniline (Michler's base) 101-61-1																					EPA-B2 IARC-2B MAK-2 NTP-R
4,4'-Methylene bis-(2-methylaniline) 838-88-0														Skin							IARC-2B MAK-2
N,N'-Methylene-bis-(5-methyloxazolidine) 66204-44-2														Sh							
Methyl ethyl ketone (MEK; 2-Butanone) 78-93-3	200	590	300	885	200	590			200	590	300	885	200	600	I (1)						EPA-I
		BEI											Skin; C								
Methyl ethyl ketone peroxide 1338-23-4		C 0.2	C 1.5							C 0.2	C 1.5										
Methyl ethyl ketoxime (2-Butanone oxime) 96-29-7														Skin; Sh				10	DSEN		MAK-2

SUBSTANCE / CAS#	ACGIH® TLVs® TWA ppm	ACGIH® TLVs® TWA mg/m³	ACGIH® TLVs® STEL/CEIL(C) ppm	ACGIH® TLVs® STEL/CEIL(C) mg/m³	OSHA PELs TWA ppm	OSHA PELs TWA mg/m³	OSHA PELs STEL/CEIL(C) ppm	OSHA PELs STEL/CEIL(C) mg/m³	NIOSH RELs TWA ppm	NIOSH RELs TWA mg/m³	NIOSH RELs STEL/CEIL(C) ppm	NIOSH RELs STEL/CEIL(C) mg/m³	DFG MAKs TWA ppm	DFG MAKs TWA mg/m³	DFG MAKs PEAK/CEIL(C) ppm	DFG MAKs PEAK/CEIL(C) mg/m³	AIHA WEELs TWA ppm	AIHA WEELs TWA mg/m³	AIHA WEELs STEL/CEIL(C) ppm	AIHA WEELs STEL/CEIL(C) mg/m³	CARCINOGENICITY CATEGORY
Methyleugenol 93-15-2																					IARC-2B NTP-R
Methyl formate (Formic acid methyl ester) 107-31-3	50	123	100	245	100	250			100	250	150	375	50	120	II (4)						
			Skin											Skin; C							
2-Methylfluoranthene 33543-31-6																					IARC-3
3-Methylfluoranthene 1706-01-0																					IARC-3
Methylglyoxal 78-98-8																					IARC-3
Methylhydrazine (Monomethyl hydrazine) 60-34-4	0.01	0.019					C 0.2	C 0.35			C 0.04*	C 0.08*									NIOSH-Ca TLV-A3
			Skin				Skin				*120-min *See* Pocket Guide App. A			Skin; Sh						
2-Methylimidazole 693-98-1																					IARC-2B
4-Methylimidazole 822-36-6																					IARC-2B
Methyl iodide 74-88-4	2	12			5	28			2	10											IARC-3 MAK-2 NIOSH-Ca
			Skin				Skin			Skin *See* Pocket Guide App. A				Skin						

SUBSTANCE / CAS#	ACGIH® TLVs® TWA ppm	mg/m³	STEL/CEIL(C) ppm	mg/m³	OSHA PELs TWA ppm	mg/m³	STEL/CEIL(C) ppm	mg/m³	NIOSH RELs TWA ppm	mg/m³	STEL/CEIL(C) ppm	mg/m³	DFG MAKs TWA ppm	mg/m³	PEAK/CEIL(C) ppm	mg/m³	AIHA WEELs TWA ppm	mg/m³	STEL/CEIL(C) ppm	mg/m³	CARCINOGENICITY CATEGORY
Methyl isoamyl ketone (Methyl-2-hexanone) 110-12-3	20	93	50	233	100	475			50	240			10	47	I (2) D						
Methyl isobutyl carbinol (Methyl amyl alcohol; 4-Methyl-2-pentanol) 108-11-2	25	104 Skin	40	167	25	100 Skin			25	100 Skin	40	165	20	85 D	I (1)						
Methyl isobutyl ketone (Hexone) 108-10-1	20	82 BEI	75	307	100	410			50	205	75	300	20	83 Skin; C	I (2)					EPA-I IARC-2B TLV-A3
Methyl isocyanate 624-83-9	0.02	0.047 Skin; DSEN	0.06	0.14	0.02	0.05 Skin			0.02	0.05 Skin			0.01	0.024 D	I (1)					
Methyl isopropyl ketone (MIPK) 563-80-4	20	70							200	705											
2-Methyl-4-isothiazolin-3-one 2682-20-4														Sh							
Methyl mercaptan (Methanethiol) 74-93-1	0.5	0.98					C 10	C 20			C 0.5* *15-min	C 1*	0.5	1 D	II (2)					
Methyl mercury 22967-92-6					0.01* *as Hg		C 0.04*		*See* Mercury, alkyl compounds					Skin; Sh						EPA-C IARC-2B* MAK-3B *compounds
Methyl methacrylate (Methacrylic acid methyl ester) 80-62-6	50	205 DSEN	100	410	100	410			100	410			50	210 Sh; C	I (2)					EPA-NL; E IARC-3 TLV-A4

SUBSTANCE CAS#	ACGIH® TLVs® TWA ppm	mg/m³	STEL/CEIL(C) ppm	mg/m³	OSHA PELs TWA ppm	mg/m³	STEL/CEIL(C) ppm	mg/m³	NIOSH RELs TWA ppm	mg/m³	STEL/CEIL(C) ppm	mg/m³	DFG MAKs TWA ppm	mg/m³	PEAK/CEIL(C) ppm	mg/m³	AIHA WEELs TWA ppm	mg/m³	STEL/CEIL(C) ppm	mg/m³	CARCINOGENICITY CATEGORY
Methyl methanesulfonate 66-27-3																					IARC-2A NTP-R
1-Methyl naphthalene [90-12-0] and 2-Methyl naphthalene [91-57-6]	0.5	3		Skin																	EPA-I* TLV-A4 *CAS: 91-57-6 only
2-Methyl-1-nitroanthraquinone, uncertain purity 129-15-7																					IARC-2B
N-Methyl-N′-nitro-N-nitrosoguanidine (MNNG) 70-25-7																					IARC-2A NTP-R
3-(Methylnitrosamino) propionitrile (3-[N-Nitrosomethylamino]propionitrile) 60153-49-3																					IARC-2B
N-Methyl-N-nitrosourethane 615-53-2																					IARC-2B
N-Methylolacrylamide 90456-67-0																					IARC-3
Methyl parathion 298-00-0	0.02 IFV			Skin; BEI_A					0.2			Skin									IARC-3 TLV-A4
2-Methylphenanthrene 832-69-9																					IARC-3

SUBSTANCE CAS#	ACGIH® TLVs®				OSHA PELs				NIOSH RELs				DFG MAKs				AIHA WEELs				CARCINOGENICITY CATEGORY
	TWA		STEL/CEIL(C)		TWA		STEL/CEIL(C)		TWA		STEL/CEIL(C)		TWA		PEAK/CEIL(C)		TWA		STEL/CEIL(C)		
	pprn	mg/m³	ppm	mg/m³	ppm	mg/m³	ppm	mg/m³	ppm	mg/m³	ppm	mg/m³	ppm	mg/m³	ppm	mg/m³	ppm	mg/m³	ppm	mg/m³	
Methyl propyl ketone (2-Pentanone) 107-87-9			150	529	200	700			150	530											
1-Methylpyrene 2381-21-7														Skin							MAK-2
7-Methylpyrido[3,4-c]-psoralen 85878-63-3																					IARC-3
N-Methyl-2-pyrrolidone 872-50-4													20	82 Vapor Skin; C	II (2)		10	Skin			
Methyl Red 493-52-7																					IARC-3
Methyl selenac 144-34-3																					IARC-3
Methyl silicate 681-84-5	1	6							1	6											
α-Methyl styrene 98-83-9	10	48					C 100	C 480	50	240	100	485	50	250 D	I (2)						IARC-2B TLV-A3
Methyltetrahydrophthalic anhydride 11070-44-3														Sa							

SUBSTANCE / CAS#	ACGIH® TLVs® TWA ppm	mg/m³	STEL/CEIL(C) ppm	mg/m³	OSHA PELs TWA ppm	mg/m³	STEL/CEIL(C) ppm	mg/m³	NIOSH RELs TWA ppm	mg/m³	STEL/CEIL(C) ppm	mg/m³	DFG MAKs TWA ppm	mg/m³	PEAK/CEIL(C) ppm	mg/m³	AIHA WEELs TWA ppm	mg/m³	STEL/CEIL(C) ppm	mg/m³	CARCINOGENICITY CATEGORY
Methylthiouracil 56-04-2																					IARC-2B
Methyltin tris(isooctyl-mercaptoacetate) [54849-38-6], **bis[Methyltin di(isooctylmercaptoacetate)] sulfide**, and **bis[Methyltin di(2-mercaptoethyloleate)] sulfide** [59118-99-9]**													0.2	1*	I (1)						
													*can also occur as vapor and aerosol **for methyltin cmpds whose organic ligands are already designated "Sa" or "Sh", these designations also apply								
Methyltrichlorosilane 75-79-6																			C 1		
Methyl vinyl ether 107-25-5													200	480	II (2) C						
Methyl vinyl ketone (3-Buten-2-one) 78-94-4			C 0.2	C 0.6 Skin; SEN											Skin; Sh						
Metolachlor 51218-45-2																					EPA-C
Metribuzin 21087-64-9	5								5												EPA-D TLV-A4

SUBSTANCE / CAS#	ACGIH® TLVs® TWA ppm	TWA mg/m³	STEL/CEIL(C) ppm	STEL/CEIL(C) mg/m³	OSHA PELs TWA ppm	TWA mg/m³	STEL/CEIL(C) ppm	mg/m³	NIOSH RELs TWA ppm	mg/m³	STEL/CEIL(C) ppm	mg/m³	DFG MAKs TWA ppm	mg/m³	PEAK/CEIL(C) ppm	mg/m³	AIHA WEELs TWA ppm	mg/m³	STEL/CEIL(C) ppm	mg/m³	CARCINOGENICITY CATEGORY
Metronidazole 443-48-1																					IARC-2B NTP-R
Mevinphos 7786-34-7	0.01 **IFV** Skin; BEI_A				0.1 Skin				0.01 Skin	0.1	0.03	0.3	0.01 Skin	0.093	II (2)						TLV-A4
Mica 12001-26-2		3 **R**			20 mppcf < 1% Crystalline silica				3* *Respirable dust; containing < 1% Quartz												
Michler's ketone 90-94-8																					IARC-2B MAK-2 NTP-R
Microbial rennets: endothiapepsin and mucorpepsin													Sa								
Microcystin-LR 101043-37-2																					IARC-2B
Mineral oil, Pure, highly and severely refined		5 **I** excluding Metal working fluids																			IARC-3* TLV-A4 *highly refined
Mineral oil, Poorly and mildly refined		excluding Metal working fluids **L**																			IARC-1* NTP-K* TLV-A2 *untreated, mildly treated
Mineral wool fiber		*See* Synthetic vitreous fibers							5* *Total Mineral wool dust, or 3 f/cc TWA (fibers ≤ 3.5 μm diam; ≥ 10 μm length)												

SUBSTANCE	ACGIH® TLVs®				OSHA PELs				NIOSH RELs				DFG MAKs				AIHA WEELs				CARCINOGENICITY
	TWA		STEL/CEIL(C)		TWA		STEL/CEIL(C)		TWA		STEL/CEIL(C)		TWA		PEAK/CEIL(C)		TWA		STEL/CEIL(C)		CATEGORY
CAS#	ppm	mg/m³	ppm	mg/m³	ppm	mg/m³	ppm	mg/m³	ppm	mg/m³	ppm	mg/m³	ppm	mg/m³	ppm	mg/m³	ppm	mg/m³	ppm	mg/m³	
Mirex 2385-85-5																					IARC-2B NTP-R
Mitomycin C 50-07-7																					IARC-2B
Mitoxantrone 65271-80-9																					IARC-2B
Modacrylic fibers																					IARC-3
Molybdenum [7439-98-7] **and insoluble compounds, as Mo**		10 **I** 3 **R**				*15 *Total dust															
Molybdenum [7439-98-7] **and soluble compounds, as Mo**		0.5 **R**				5															TLV-A3
Molybdenum trioxide 1313-27-5																					MAK-3B
Monochloroacetic acid 79-11-8	0.5 **IFV**		2 **IFV** Skin														0.5		Skin		TLV-A4
3-Monochloro-1,2-pro-panediol 96-24-2													0.005	0.025	II (8) Skin; D						IARC-2B MAK-3B

SUBSTANCE / CAS#	ACGIH® TLVs® TWA ppm	ACGIH® TLVs® TWA mg/m³	ACGIH® TLVs® STEL/CEIL(C) ppm	ACGIH® TLVs® STEL/CEIL(C) mg/m³	OSHA PELs TWA ppm	OSHA PELs TWA mg/m³	OSHA PELs STEL/CEIL(C) ppm	OSHA PELs STEL/CEIL(C) mg/m³	NIOSH RELs TWA ppm	NIOSH RELs TWA mg/m³	NIOSH RELs STEL/CEIL(C) ppm	NIOSH RELs STEL/CEIL(C) mg/m³	DFG MAKs TWA ppm	DFG MAKs TWA mg/m³	DFG MAKs PEAK/CEIL(C) ppm	DFG MAKs PEAK/CEIL(C) mg/m³	AIHA WEELs TWA ppm	AIHA WEELs TWA mg/m³	AIHA WEELs STEL/CEIL(C) ppm	AIHA WEELs STEL/CEIL(C) mg/m³	CARCINOGENICITY CATEGORY
Monocrotaline 315-22-0																					IARC-2B
Monocrotophos 6923-22-4	0.05 **IFV** Skin; BEI$_A$									0.25											TLV-A4
Monomethyltin compounds, as Sn, except **Methyltin tris(iso-octylmercaptoacetate)** [54849-38-6], **bis[Methyltin di(isooctylmercaptoacetate)] sulfide, and bis[Methyltin di(2-mercaptoethyloleate)] sulfide** [59118-99-9]													0.004	0.02*	I (1)						MAK-4
Mono-n-butyltin compounds, as Sn													0.004	0.02* Skin**; C	I (1)						MAK-4
Mono-n-octyltin compounds, as Sn													0.002	0.0098* Skin**; C	II (2)						MAK-4

Monomethyltin compounds notes:
*can also occur as vapor and aerosol
**for methyltin cmpds whose organic ligands are already designated "Sa" or "Sh", these designations also apply

C

Mono-n-butyltin compounds notes:
*can also be found as vapor
**for n-butyltin cmpds whose organic ligands are already designated "Sa" or "Sh", these designations also apply

Mono-n-octyltin compounds notes:
*can also be found as vapor
**for n-octyltin cmpds whose organic ligands are already designated "Sa" or "Sh", these designations also apply

SUBSTANCE CAS#	ACGIH TLVs TWA ppm	TWA mg/m³	STEL/CEIL(C) ppm	STEL/CEIL(C) mg/m³	OSHA PELs TWA ppm	TWA mg/m³	STEL/CEIL(C) ppm	STEL/CEIL(C) mg/m³	NIOSH RELs TWA ppm	TWA mg/m³	STEL/CEIL(C) ppm	STEL/CEIL(C) mg/m³	DFG MAKs TWA ppm	TWA mg/m³	PEAK/CEIL(C) ppm	PEAK/CEIL(C) mg/m³	AIHA WEELs TWA ppm	TWA mg/m³	STEL/CEIL(C) ppm	STEL/CEIL(C) mg/m³	CARCINOGENICITY CATEGORY
Monuron 150-68-5																					IARC-3
Morpholine 110-91-8	20	71		Skin	20	70		Skin	20	70	30	105 Skin	10	36	I (2)	D					IARC-3 TLV-A4
2-(4-Morpholinylmer-capto)benzothiazole 102-77-2															Sh						
Musk ambrette 83-66-9																					IARC-3
Musk xylene 81-15-2																					IARC-3
Mustard gas (2,2′-Dichlorodiethyl sulfide) 505-60-2															Skin						IARC-1 MAK-1 NTP-K
Nafenopin 3771-19-5																					IARC-2B
Naled (Dibrom; Dimethyl-1,2-dibromo-2,2-dichloroeth-ylphosphate) 300-76-5	0.1 **IFV**			Skin; DSEN; BEI$_A$	3				3			Skin	1 **I**		II (2)	Skin; Sh; C					TLV-A4
Naphtha, coal tar 8030-30-6					100	400			100	400											

SUBSTANCE / CAS#	ACGIH® TLVs® TWA ppm	mg/m³	STEL/CEIL(C) ppm	mg/m³	OSHA PELs TWA ppm	mg/m³	STEL/CEIL(C) ppm	mg/m³	NIOSH RELs TWA ppm	mg/m³	STEL/CEIL(C) ppm	mg/m³	DFG MAKs TWA ppm	mg/m³	PEAK/CEIL(C) ppm	mg/m³	AIHA WEELs TWA ppm	mg/m³	STEL/CEIL(C) ppm	mg/m³	CARCINOGENICITY CATEGORY
Naphtha, petroleum, hydrotreated, heavy 64742-48-9													50	300	II (2) D						
Naphthalene 91-20-3	10	52		Skin; BEI	10	50			10	50	15	75			Skin; 3B						EPA-CBD; NTP-R C TLV-A3 IARC-2B MAK-2
1,5-Naphthalenediamine 2243-62-1															Skin; Sh						IARC-3 MAK-2
1,5-Naphthalene diisocyanate (NDI) 3173-72-6									0.005	0.04	C 0.02* *10-min	C 0.17*			Sa						IARC-3 MAK-3B
1,8-Naphthalic anhydride 81-84-5															Sh						
Naphthenic acids [1338-24-5] and Na-, Ca-, K- [61789-36-4; 61790-13-4; 66072-08-0] naphthenates																					MAK-3B
Naphtho[1,2-b]fluor-anthene 111189-32-3																					IARC-3
Naphtho[2,1-a]fluor-anthene 203-20-3																					IARC-3

SUBSTANCE / CAS#	ACGIH® TLVs® TWA ppm	TWA mg/m³	STEL/CEIL(C) ppm	mg/m³	OSHA PELs TWA ppm	mg/m³	STEL/CEIL(C) ppm	mg/m³	NIOSH RELs TWA ppm	mg/m³	STEL/CEIL(C) ppm	mg/m³	DFG MAKs TWA ppm	mg/m³	PEAK/CEIL(C) ppm	mg/m³	AIHA WEELs TWA ppm	mg/m³	STEL/CEIL(C) ppm	mg/m³	CARCINOGENICITY CATEGORY
Naphtho[2,3-e]pyrene 193-09-9																					IARC-3
α-Naphthylamine (1-Naphthylamine) 134-32-7					*See* 29 CFR 1910.1003				*See* Pocket Guide App. A												IARC-3 NIOSH-Ca OSHA-Ca
β-Naphthylamine (2-Naphthylamine) 91-59-8		L			*See* 29 CFR 1910.1003				*See* Pocket Guide App. A				can also occur as vapor and aerosol Skin; 3A								IARC-1　OSHA-Ca MAK-1　TLV-A1 NIOSH-Ca NTP-K
Natural gas 8006-14-2	(*See* Appendix F in *TLVs*® and *BEIs*® book) NIC-D, EX																				
Natural rubber latex, as inhalable allergenic proteins 9003-31-0; 9006-04-6	0.0001 I												Sah								
Nemalite, fibrous dust 1317-43-7																					MAK-3B
Neon 7440-01-9	*Documentation* withdrawn; *see* Appendix F in *TLVs*® and *BEIs*® book Simple asphyxiant(D)																				
NIAX® Catalyst ESN (Dimethylaminopropionitrile/ bis[2-dimethylamino]ethyl ether mixture) 62765-93-9									minimize exposure *See* Pocket Guide App. C												

SUBSTANCE CAS#	ACGIH® TLVs®				OSHA PELs				NIOSH RELs				DFG MAKs				AIHA WEELs				CARCINOGENICITY CATEGORY
	TWA		STEL/CEIL(C)		TWA		STEL/CEIL(C)		TWA		STEL/CEIL(C)		TWA		PEAK/CEIL(C)		TWA		STEL/CEIL(C)		
	ppm	mg/m³	ppm	mg/m³	ppm	mg/m³	ppm	mg/m³	ppm	mg/m³	ppm	mg/m³	ppm	mg/m³	ppm	mg/m³	ppm	mg/m³	ppm	mg/m³	
Nickel, alloys													for Nickel alloys containing bio-available Nickel, *see* Nickel and Nickel compounds								IARC-2B
Nickel [7440-02-0] **compounds**										0.015* *as Ni *See* Pocket Guide App. A			There is sufficient evidence of sensitizing effects on the respiratory tract only for water-soluble Nickel cmpds Inhalable fraction Sah								IARC-1 MAK-1 NIOSH-Ca NTP-K
Nickel, elemental 7440-02-0		1.5 I				1				0.015* *as Ni *See* Pocket Guide App. A			There is sufficient evidence of sensitizing effects on the respiratory tract only for water-soluble Nickel cmpds Inhalable fraction Sah								IARC-2B MAK-1 NIOSH-Ca NTP-K TLV-A5
Nickel, insoluble **compounds, as Ni** 7440-02-0 *Inorganic only		0.2 I*				1				0.015* *as Ni *See* Pocket Guide App. A											NIOSH-Ca NTP-K TLV-A1
Nickel, soluble **compounds, as Ni** 7440-02-0 *Inorganic only		0.1 I*				1				0.015* *as Ni *See* Pocket Guide App. A											NIOSH-Ca NTP-K TLV-A4
Nickel acetate 373-02-4										0.015* *as Ni *See* Pocket Guide App. A			and similar soluble salts Inhalable fraction Sah								MAK-1 NIOSH-Ca NTP-K
Nickel carbonate 3333-67-3										0.015* *as Ni *See* Pocket Guide App. A			Inhalable fraction Sah								IARC-1 MAK-1 NIOSH-Ca NTP-K

SUBSTANCE / CAS#	ACGIH® TLVs® TWA ppm	TWA mg/m³	STEL/CEIL(C) ppm	STEL/CEIL(C) mg/m³	OSHA PELs TWA ppm	TWA mg/m³	STEL/CEIL(C) ppm	STEL/CEIL(C) mg/m³	NIOSH RELs TWA ppm	TWA mg/m³	STEL/CEIL(C) ppm	STEL/CEIL(C) mg/m³	DFG MAKs TWA ppm	TWA mg/m³	PEAK/CEIL(C) ppm	PEAK/CEIL(C) mg/m³	AIHA WEELs TWA ppm	TWA mg/m³	STEL/CEIL(C) ppm	STEL/CEIL(C) mg/m³	CARCINOGENICITY CATEGORY
Nickel carbonyl, as Ni 13463-39-3			C 0.05	C 0.12	0.001	0.007			0.001	0.007 *See* Pocket Guide App. A											EPA-B2 TLV-A3 IARC-1 NIOSH-Ca NTP-K
Nickel chloride 7718-54-9		*See* Nickel, soluble compounds, as Ni				1* *as Ni				0.015* *as Ni *See* Pocket Guide App. A				Inhalable fraction Sah							IARC-1 MAK-1 NIOSH-Ca NTP-K
Nickel dioxide 12035-36-8										0.015* *as Ni *See* Pocket Guide App. A				Inhalable fraction Sah							IARC-1 MAK-1 NIOSH-Ca NTP-K
Nickel hydroxide 12054-48-7										0.015* *as Ni *See* Pocket Guide App. A				Inhalable fraction Sah							IARC-1 MAK-1 NIOSH-Ca NTP-K
Nickel oxide 1313-99-1		*See* Nickel, insoluble compounds, as Ni				1* *as Ni				0.015* *as Ni *See* Pocket Guide App. A				Inhalable fraction Sah							IARC-1 MAK-1 NIOSH-Ca NTP-K
Nickel refinery dust																					EPA-A
Nickel sesquioxide 1314-06-3		*See* Nickel, insoluble compounds, as Ni				1* *as Ni				0.015* *as Ni *See* Pocket Guide App. A				Inhalable fraction Sah							IARC-1 MAK-1 NIOSH-Ca NTP-K
Nickel subsulfide 12035-72-2		*See* Nickel, soluble compounds, as Ni								0.015* *as Ni *See* Pocket Guide App. A				Inhalable fraction Sah							EPA-A NTP-K IARC-1 MAK-1 NIOSH-Ca
Nickel sulfate 7786-81-4		*See* Nickel, soluble compounds, as Ni				1* *as Ni				0.015* *as Ni *See* Pocket Guide App. A				Inhalable fraction Sah							IARC-1 MAK-1 NIOSH-Ca NTP-K

SUBSTANCE / CAS#	ACGIH® TLVs TWA ppm	ACGIH® TLVs TWA mg/m³	ACGIH® TLVs STEL/CEIL(C) ppm	ACGIH® TLVs STEL/CEIL(C) mg/m³	OSHA PELs TWA ppm	OSHA PELs TWA mg/m³	OSHA PELs STEL/CEIL(C) ppm	OSHA PELs STEL/CEIL(C) mg/m³	NIOSH RELs TWA ppm	NIOSH RELs TWA mg/m³	NIOSH RELs STEL/CEIL(C) ppm	NIOSH RELs STEL/CEIL(C) mg/m³	DFG MAKs TWA ppm	DFG MAKs TWA mg/m³	DFG MAKs PEAK/CEIL(C) ppm	DFG MAKs PEAK/CEIL(C) mg/m³	AIHA WEELs TWA ppm	AIHA WEELs TWA mg/m³	AIHA WEELs STEL/CEIL(C) ppm	AIHA WEELs STEL/CEIL(C) mg/m³	CARCINOGENICITY CATEGORY
Nickel sulfide 16812-54-7														Inhalable fraction Sah							IARC-1 MAK-1 NTP-K
Nicotine 54-11-5		0.5 Skin				0.5 Skin				0.5 Skin				Skin							
Nifurthiazole (2-[2-Form-ylhydrazino]-4-[5-nitro-2-furyl]thiazole) 3570-75-0																					IARC-2B
Niridazole 61-57-4																					IARC-2B
Nithiazide 139-94-6																					IARC-3
Nitrapyrin (2-Chloro-6-(trichloromethyl) pyridine) 1929-82-4		10		20	15*; 5**				10*; 5**		20*										TLV-A4
					*Total dust **Respirable fraction				*Total dust **Respirable fraction												
Nitric acid 7697-37-2	2	5.2	4	10	2	5			2	5	4	10									
Nitric oxide 10102-43-9	25	31 BEI_M			25	30			25	30			0.5	0.63 D	I (2)						
Nitrilotriacetic acid 139-13-9														avoid simultaneous exposure to Iron compounds							IARC-2B* MAK-3A** NTP-R *and its salts **and its sodium salts

SUBSTANCE CAS#	ACGIH® TLVs® TWA ppm	mg/m³	STEL/CEIL(C) ppm	mg/m³	OSHA PELs TWA ppm	mg/m³	STEL/CEIL(C) ppm	mg/m³	NIOSH RELs TWA ppm	mg/m³	STEL/CEIL(C) ppm	mg/m³	DFG MAKs TWA ppm	mg/m³	PEAK/CEIL(C) ppm	mg/m³	AIHA WEELs TWA ppm	mg/m³	STEL/CEIL(C) ppm	mg/m³	CARCINOGENICITY CATEGORY
5-Nitroacenaphthene 602-87-9																					IARC-2B MAK-2
4-Nitro-4'-aminodi-phenylamine-2-sulfonic acid 91-29-2													Sh								
2-Nitro-4-aminophenol (4-Amino-2-nitrophenol) 119-34-6													Skin								IARC-3 MAK-3B
Nitroaniline, p-isomer 100-01-6	3 Skin; BEI_M				1 Skin	6			3 Skin				Skin								MAK-3A TLV-A4
5-Nitro-o-anisidine 99-59-2																					IARC-3
2-Nitroanisole 91-23-6																					IARC-2B MAK-2 NTP-R
9-Nitroanthracene 602-60-8																					IARC-3
7-Nitrobenz[a]anthra-cene 20268-51-3																					IARC-3
3-Nitrobenzanthrone 17117-34-9																					IARC-2B

SUBSTANCE / CAS#	ACGIH® TLVs® TWA ppm	mg/m³	STEL/CEIL(C) ppm	mg/m³	OSHA PELs TWA ppm	mg/m³	STEL/CEIL(C) ppm	mg/m³	NIOSH RELs TWA ppm	mg/m³	STEL/CEIL(C) ppm	mg/m³	DFG MAKs TWA ppm	mg/m³	PEAK/CEIL(C) ppm	mg/m³	AIHA WEELs TWA ppm	mg/m³	STEL/CEIL(C) ppm	mg/m³	CARCINOGENICITY CATEGORY
Nitrobenzene 98-95-3	1	5			1	5			1	5			0.1	0.51*	II (4)						EPA-L NTP-R IARC-2B TLV-A3 MAK-4
		Skin				Skin				Skin			*can also occur as vapor and aerosol Skin; C								
6-Nitrobenzo[a]pyrene 63041-90-7																					IARC-3
3-Nitrobenzoic acid 121-92-6														Skin							
4-Nitrobenzoic acid 62-23-7													1 I		I (2)						MAK-3B
													D								
4-(2-Nitrobutyl)morpholine (70% w/v) [2224-44-4] **and 4,4′-(2-Ethyl-2-nitro-1,3-propanediyl)bismorpholine (20% w/v)** [1854-23-5] **mixture**													0.5	4.2	I (2)						
													Sh; D								
Nitrochlorobenzene, m-isomer (1-Chloronitro-benzene) 121-73-3														Skin							IARC-3
Nitrochlorobenzene, o-isomer (2-Chloronitro-benzene) 88-73-3														Skin							IARC-3 MAK-3B
Nitrochlorobenzene, p-isomer (4-Chloronitro-benzene) 100-00-5	0.1	0.64			1									Skin							IARC-3 MAK-3B NIOSH-Ca TLV-A3
		Skin; BEI_M				Skin			Skin See Pocket Guide App. A												

SUBSTANCE / CAS#	ACGIH® TLVs® TWA ppm	mg/m³	STEL/CEIL(C) ppm	mg/m³	OSHA PELs TWA ppm	mg/m³	STEL/CEIL(C) ppm	mg/m³	NIOSH RELs TWA ppm	mg/m³	STEL/CEIL(C) ppm	mg/m³	DFG MAKs TWA ppm	mg/m³	PEAK/CEIL(C) ppm	mg/m³	AIHA WEELs TWA ppm	mg/m³	STEL/CEIL(C) ppm	mg/m³	CARCINOGENICITY CATEGORY
6-Nitrochrysene 7496-02-8																					IARC-2A NTP-R
Nitrocumene, p-isomer 1817-47-6														Sh							
4-Nitrodiphenyl (4-Nitrobiphenyl) 92-93-3	Skin; L				See 29 CFR 1910.1003				See Pocket Guide App. A				Skin								IARC-3 TLV-A2 MAK-2 NIOSH-Ca OSHA-Ca
Nitroethane 79-24-3	100	307			100	310			100	310			10	31	II (4)						Skin; D
Nitrofen 1836-75-5																					IARC-2B* NTP-R *technical grade
3-Nitrofluoranthene 892-21-7																					IARC-3
2-Nitrofluorene 607-57-8																					IARC-2B
Nitrofural (Nitrofurazone) 59-87-0																					IARC-3
Nitrofurantoin 67-20-9																					IARC-3

SUBSTANCE / CAS#	ACGIH® TLVs® TWA ppm	mg/m³	STEL/CEIL(C) ppm	mg/m³	OSHA PELs TWA ppm	mg/m³	STEL/CEIL(C) ppm	mg/m³	NIOSH RELs TWA ppm	mg/m³	STEL/CEIL(C) ppm	mg/m³	DFG MAKs TWA ppm	mg/m³	PEAK/CEIL(C) ppm	mg/m³	AIHA WEELs TWA ppm	mg/m³	STEL/CEIL(C) ppm	mg/m³	CARCINOGENICITY CATEGORY
1-[(5-Nitrofurfurylidene)amino]-2-imidazolidinone 555-84-0																					IARC-2B
Nitrogen 7440-01-9	*Documentation withdrawn; see* Appendix F in *TLVs® and BEIs®* book Simple asphyxiant (D)																				
Nitrogen dioxide 10102-44-0	0.2	0.38					C 5	C 9			1	1.8	0.5	0.95	I (1) D						MAK-3B TLV-A4
Nitrogen mustard (N-Methyl-bis[2-chloroethyl]amine) 51-75-2															Skin; Sh; 2						IARC-2A MAK-1
Nitrogen mustard hydrochloride (Mechlorethamine hydrochloride) 55-86-7																					NTP-R
Nitrogen mustard N-oxide 126-85-2																					IARC-2B
Nitrogen trifluoride 7783-54-2	10	29 BEI_M			10	29			10	29											
Nitroglycerin (NG) 55-63-0	0.05	0.46 Skin				Skin	C 0.2	C 2		Skin		0.1	0.01	0.094	II (1) Skin; C						MAK-3B
Nitroguanidine 556-88-7																					EPA-D

SUBSTANCE CAS#	ACGIH® TLVs® TWA ppm	mg/m³	STEL/CEIL(C) ppm	mg/m³	OSHA PELs TWA ppm	mg/m³	STEL/CEIL(C) ppm	mg/m³	NIOSH RELs TWA ppm	mg/m³	STEL/CEIL(C) ppm	mg/m³	DFG MAKs TWA ppm	mg/m³	PEAK/CEIL(C) ppm	mg/m³	AIHA WEELs TWA ppm	mg/m³	STEL/CEIL(C) ppm	mg/m³	CARCINOGENICITY CATEGORY
Nitromethane 75-52-5	20	50			100	250									Skin						IARC-2B MAK-3B NTP-R TLV-A3
1-Nitronaphthalene 86-57-7																					IARC-3 MAK-3B
2-Nitronaphthalene 581-89-5									See Pocket Guide App. A												IARC-3 NIOSH-Ca* MAK-2 *since metabolized to β-Naphthylamine
2-Nitro-p-phenylene-diamine (1,4-Diamino-2-nitrobenzene) 5307-14-2															Skin; Sh						IARC-3 MAK-3B
3-Nitroperylene 20589-63-3																					IARC-3
1-Nitropropane 108-03-2	25	91			25	90			25	90			2	7.4	I (8)						TLV-A4
													See 2-Nitropropane when measurably contaminated with that isomer Skin; D								
2-Nitropropane 79-46-9	10	36			25	90			See Pocket Guide App. A						Skin						IARC-2B NTP-R MAK-2 TLV-A3 NIOSH-Ca
Nitropyrenes 789-07-1; 5522-43-0; 28767-61-5; 57835-92-4; 63021-86-3; 75321-19-6; 78432-19-6																					IARC-2A*; 2B**, 3*** MAK-3B**** NTP-R*, ** *1-Nitropyrene **4-Nitropyrene ***2-Nitropyrene ****mono-, di-, tri-, tetra-isomers

SUBSTANCE / CAS#	ACGIH® TLVs® TWA ppm	mg/m³	STEL/CEIL(C) ppm	mg/m³	OSHA PELs TWA ppm	mg/m³	STEL/CEIL(C) ppm	mg/m³	NIOSH RELs TWA ppm	mg/m³	STEL/CEIL(C) ppm	mg/m³	DFG MAKs TWA ppm	mg/m³	PEAK/CEIL(C) ppm	mg/m³	AIHA WEELs TWA ppm	mg/m³	STEL/CEIL(C) ppm	mg/m³	CARCINOGENICITY CATEGORY
N'-Nitrosoanabasine (NAB) 37620-20-5																					IARC-3
N'-Nitrosoanatabine (NAT) 71267-22-6																					IARC-3
N-Nitrosodi-n-butyl-amine (DBN) 924-16-3														Skin							EPA-B2 IARC-2B MAK-2 NTP-R
N-Nitrosodiethanolamine (NDELA) 1116-54-7														Skin							EPA-B2 IARC-2B MAK-2 NTP-R
N-Nitrosodiethylamine (NDEA) 55-18-5														Skin							EPA-B2 IARC-2A MAK-2 NTP-R
N-Nitrosodiisopropyl-amine 601-77-4														Skin							MAK-2
N-Nitrosodimethylamine (N,N-Dimethylnitrosoamine) 62-75-9	Skin; L				*See* 29 CFR 1910.1003				*See* Pocket Guide App. A					Skin							EPA-B2 NTP-R IARC-2A OSHA-Ca MAK-2 TLV-A3 NIOSH-Ca
N-Nitrosodiphenyl-amine 86-30-6																					EPA-B2 IARC-3 MAK-3B
p-Nitrosodiphenyl-amine 156-10-5																					IARC-3

SUBSTANCE CAS#	ACGIH® TLVs® TWA ppm	ACGIH® TLVs® TWA mg/m³	ACGIH® TLVs® STEL/CEIL(C) ppm	ACGIH® TLVs® STEL/CEIL(C) mg/m³	OSHA PELs TWA ppm	OSHA PELs TWA mg/m³	OSHA PELs STEL/CEIL(C) ppm	OSHA PELs STEL/CEIL(C) mg/m³	NIOSH RELs TWA ppm	NIOSH RELs TWA mg/m³	NIOSH RELs STEL/CEIL(C) ppm	NIOSH RELs STEL/CEIL(C) mg/m³	DFG MAKs TWA ppm	DFG MAKs TWA mg/m³	DFG MAKs PEAK/CEIL(C) ppm	DFG MAKs PEAK/CEIL(C) mg/m³	AIHA WEELs TWA ppm	AIHA WEELs TWA mg/m³	AIHA WEELs STEL/CEIL(C) ppm	AIHA WEELs STEL/CEIL(C) mg/m³	CARCINOGENICITY CATEGORY
N-Nitrosodi-n-propyl-amine (NDPA) 621-64-7														Skin							EPA-B2 IARC-2B MAK-2 NTP-R
N-Nitrosoethylphenyl-amine 612-64-6														Skin							MAK-2
N-Nitroso-N-ethylurea 759-73-9																					IARC-2A NTP-R
N-Nitrosofolic acid 29291-35-8																					IARC-3
N-Nitrosoguvacine 55557-01-2																					IARC-3
N-Nitrosoguvacoline 55557-02-3																					IARC-3
N-Nitrosohydroxy-proline 30310-80-6																					IARC-3
3-(N-Nitrosomethyl-amino)propion-aldehyde 85502-23-4																					IARC-3
4-(N-Nitrosomethyl-amino)-4-(3-pyridyl)-1-butanal (NNA) 16543-55-8																					IARC-1

SUBSTANCE CAS#	ACGIH® TLVs® TWA ppm	ACGIH® TLVs® TWA mg/m³	ACGIH® TLVs® STEL/CEIL(C) ppm	ACGIH® TLVs® STEL/CEIL(C) mg/m³	OSHA PELs TWA ppm	OSHA PELs TWA mg/m³	OSHA PELs STEL/CEIL(C) ppm	OSHA PELs STEL/CEIL(C) mg/m³	NIOSH RELs TWA ppm	NIOSH RELs TWA mg/m³	NIOSH RELs STEL/CEIL(C) ppm	NIOSH RELs STEL/CEIL(C) mg/m³	DFG MAKs TWA ppm	DFG MAKs TWA mg/m³	DFG MAKs PEAK/CEIL(C) ppm	DFG MAKs PEAK/CEIL(C) mg/m³	AIHA WEELs TWA ppm	AIHA WEELs TWA mg/m³	AIHA WEELs STEL/CEIL(C) ppm	AIHA WEELs STEL/CEIL(C) mg/m³	CARCINOGENICITY CATEGORY
4-(N-Nitrosomethyl-amino)-1-(3-pyridyl)-1-butanone 64091-91-4																					IARC-1 NTP-R
N-Nitrosomethylethyl-amine 10595-95-6														Skin							EPA-B2 IARC-2B MAK-2
N-Nitrosomethyl-phenylamine 614-00-6														Skin							MAK-2
N-Nitroso-N-methylurea 684-93-5																					IARC-2A NTP-R
N-Nitrosomethylvinyl-amine 4549-40-0																					IARC-2B NTP-R
N-Nitrosomorpholine (NMOR) 59-89-2														Skin							IARC-2B MAK-2 NTP-R
N'-Nitrosonornicotine (NNN) 80508-23-2																					IARC-1 NTP-R
N-Nitrosopiperidine (NPIP) 100-75-4														Skin							IARC-2B MAK-2 NTP-R
N-Nitrosoproline 7519-36-0																					IARC-3

SUBSTANCE / CAS#	ACGIH® TLVs® TWA ppm	mg/m³	STEL/CEIL(C) ppm	mg/m³	OSHA PELs TWA ppm	mg/m³	STEL/CEIL(C) ppm	mg/m³	NIOSH RELs TWA ppm	mg/m³	STEL/CEIL(C) ppm	mg/m³	DFG MAKs TWA ppm	mg/m³	PEAK/CEIL(C) ppm	mg/m³	AIHA WEELs TWA ppm	mg/m³	STEL/CEIL(C) ppm	mg/m³	CARCINOGENICITY CATEGORY
N-Nitrosopyrrolidine (NPYR) 930-55-2														Skin							EPA-B2 IARC-2B MAK-2 NTP-R
N-Nitrososarcosine 13256-22-9																					IARC-2B NTP-R
Nitrotoluene, m-isomer (3-Nitrotoluene) 99-08-1	2	11 Skin; BEI_M			5	30 Skin			2	11 Skin				Skin							IARC-3 MAK-3B
Nitrotoluene, o-isomer (2-Nitrotoluene) 88-72-2	2	11 Skin; BEI_M			5	30 Skin			2	11 Skin				Skin; 3B							IARC-2A MAK-2 NTP-R
Nitrotoluene, p-isomer (4-Nitrotoluene) 99-99-0	2	11 Skin; BEI_M			5	30 Skin			2	11 Skin				Skin							IARC-3 MAK-3B
5-Nitro-o-toluidine (4-Nitro-2-aminotoluene) 99-55-8		1 I																			IARC-3 MAK-2 TLV-A3
3-Nitro-1,2,4-triazol-5-one 932-64-9																		2* *OARS WEEL			
Nitrous oxide 10024-97-2	50	90							25* *over the time exposed; for exposure to waste anesthetic gases	46*			100	180	II (2) C						TLV-A4
Nitrovin 804-36-4																					IARC-3

SUBSTANCE / CAS#	ACGIH® TLVs® TWA ppm	mg/m³	STEL/CEIL(C) ppm	mg/m³	OSHA PELs TWA ppm	mg/m³	STEL/CEIL(C) ppm	mg/m³	NIOSH RELs TWA ppm	mg/m³	STEL/CEIL(C) ppm	mg/m³	DFG MAKs TWA ppm	mg/m³	PEAK/CEIL(C) ppm	mg/m³	AIHA WEELs TWA ppm	mg/m³	STEL/CEIL(C) ppm	mg/m³	CARCINOGENICITY CATEGORY
Nodularins 118399-22-7																					IARC-3
Nonane 111-84-2	200	1050							200	1050											
Nonane, all isomers 111-84-2; 3522-94-9	TLV® withdrawn; *see* Nonane																				
Nonabromodiphenyl ether 63936-56-1																					EPA-D
n-Nonyl mercaptan 1455-21-6											C 0.5*	C 3.3* *15-min									
Norethisterone 68-22-4																					NTP-R
Nuisance particles					*See* Particles not otherwise classified/regulated								*See* Dust, general threshold limit value								
Nylon 6 25038-54-4																					IARC-3
Oakmoss extracts															Sh						

SUBSTANCE / CAS#	ACGIH® TLVs® TWA ppm	TWA mg/m³	STEL/CEIL(C) ppm	STEL/CEIL(C) mg/m³	OSHA PELs TWA ppm	TWA mg/m³	STEL/CEIL(C) ppm	STEL/CEIL(C) mg/m³	NIOSH RELs TWA ppm	TWA mg/m³	STEL/CEIL(C) ppm	STEL/CEIL(C) mg/m³	DFG MAKs TWA ppm	TWA mg/m³	PEAK/CEIL(C) ppm	PEAK/CEIL(C) mg/m³	AIHA WEELs TWA ppm	TWA mg/m³	STEL/CEIL(C) ppm	STEL/CEIL(C) mg/m³	CARCINOGENICITY CATEGORY
Ochratoxin A 303-47-9														3B							IARC-2B MAK-2 NTP-R
Octabromodiphenyl ether 32536-52-0																					EPA-D
Octachloronaphthalene 2234-13-1		0.1		0.3		0.1				0.1		0.3									
		Skin				Skin				Skin											
Octadecyl mercaptan 2885-00-9											C 0.5*	C 5.9*									
											*15-min										
Octamethylcyclotetra-siloxane 556-67-2																	10*				
																	*OARS WEEL				
Octane, all isomers 111-65-9; 540-84-1	300	1401			500	2350			75	350	C 385*	C 1800*	500	2400	II (2)						EPA-II*
						n-Octane only				*15-min CAS: 111-65-9 only			except Trimethylpentane isomers	D							*CAS: 540-84-1 (oral)
1-Octanol 111-87-5																	50				
1-Octene 111-66-0																	75				
Octogen 2691-41-0																					EPA-D

SUBSTANCE / CAS#	ACGIH® TLVs® TWA ppm	mg/m³	STEL/CEIL(C) ppm	mg/m³	OSHA PELs TWA ppm	mg/m³	STEL/CEIL(C) ppm	mg/m³	NIOSH RELs TWA ppm	mg/m³	STEL/CEIL(C) ppm	mg/m³	DFG MAKs TWA ppm	mg/m³	PEAK/CEIL(C) ppm	mg/m³	AIHA WEELs TWA ppm	mg/m³	STEL/CEIL(C) ppm	mg/m³	CARCINOGENICITY CATEGORY
2-Octyl-4-isothiazolin-3-one 26530-20-1														0.05 I		I (2)						
													Skin; Sh; C								
n-Octyl mercaptan 111-88-6											C 0.5*	C 3*									
											*15-min										
4-tert-Octylphenol 140-66-9														0.5	4.3*	I (1)						
													*can also occur as vapor and aerosol D								
Oil mist, mineral 8012-95-1	TLV® withdrawn; see Mineral oil					5				5		10									
Oil Orange SS 2646-17-5																						IARC-2B
Olaquindox 23696-28-8														SP; 2								MAK-3B
Oleyl sarcosine 110-25-8														0.1 I		II (2)						
													D								
Orange I 523-44-4																						IARC-3
Orange G 1936-15-8																						IARC-3

SUBSTANCE / CAS#	ACGIH® TLVs® TWA ppm	TWA mg/m³	STEL/CEIL(C) ppm	STEL/CEIL(C) mg/m³	OSHA PELs TWA ppm	TWA mg/m³	STEL/CEIL(C) ppm	STEL/CEIL(C) mg/m³	NIOSH RELs TWA ppm	TWA mg/m³	STEL/CEIL(C) ppm	STEL/CEIL(C) mg/m³	DFG MAKs TWA ppm	TWA mg/m³	PEAK/CEIL(C) ppm	PEAK/CEIL(C) mg/m³	AIHA WEELs TWA ppm	TWA mg/m³	STEL/CEIL(C) ppm	STEL/CEIL(C) mg/m³	CARCINOGENICITY CATEGORY
Oryzalin 19044-88-3																					EPA-C
Osmium tetroxide 20816-12-0	0.0002	0.0016	0.0006	0.0047		0.002*			0.0002	0.002	0.0006	0.006									
						*as Os															
Oxalic acid, anhydrous and dihydrate 144-62-7		1		2		1				1		2									
Oxazepam 604-75-1																					IARC-2B
p,p'-Oxybis(benzene-sulfonyl hydrazide) 80-51-3		0.1																			
Oxygen difluoride 7783-41-7			C 0.05	C 0.11	0.05	0.1					C 0.05	C 0.1									
Oxymetholone 434-07-1																					NTP-R
Oxyphenbutazone 129-20-4																					IARC-3

SUBSTANCE / CAS#	ACGIH® TLVs® TWA ppm	mg/m³	STEL/CEIL(C) ppm	mg/m³	OSHA PELs TWA ppm	mg/m³	STEL/CEIL(C) ppm	mg/m³	NIOSH RELs TWA ppm	mg/m³	STEL/CEIL(C) ppm	mg/m³	DFG MAKs TWA ppm	mg/m³	PEAK/CEIL(C) ppm	mg/m³	AIHA WEELs TWA ppm	mg/m³	STEL/CEIL(C) ppm	mg/m³	CARCINOGENICITY CATEGORY
Ozone [10028-15-6]					0.1	0.2					C 0.1	C 0.2									MAK-3B TLV-A4
Heavy work	0.05	0.1																			
Moderate work	0.08	0.16																			
Light work	0.1	0.2																			
Light, moderate, or heavy workload	0.2	0.4	≤ 2 hours																		
Palladium chloride [7647-10-1] and other bio-available Pd(II) compounds														Sh							
Papain 9001-73-4														Sa							
Paraffin wax fume 8002-74-2		2								2											
Paraquat 4685-14-7		(0.5) (0.1 R) NIC-0.05 I			NIC-A4 as the cation NIC-Skin	0.5* *Respirable dust Skin															
Paraquat dichloride 1910-42-5		(0.5) (0.1 R) NIC-0.05 I			NIC-A4 as the cation NIC-Skin	0.5* *Respirable dust Skin				0.1* *Respirable dust Skin				0.1 I Skin		I (1)					EPA-C
Paraquat methosulfate (Paraquat dimethyl sulfate) 2074-50-2		(0.5) (0.1 R) NIC-0.05 I			NIC-A4 as the cation NIC-Skin	0.5* *Respirable dust Skin															
Parasorbic acid 10048-32-5																					IARC-3

SUBSTANCE / CAS#	ACGIH® TLVs® TWA ppm	TWA mg/m³	STEL/CEIL(C) ppm	STEL/CEIL(C) mg/m³	OSHA PELs TWA ppm	TWA mg/m³	STEL/CEIL(C) ppm	STEL/CEIL(C) mg/m³	NIOSH RELs TWA ppm	TWA mg/m³	STEL/CEIL(C) ppm	STEL/CEIL(C) mg/m³	DFG MAKs TWA ppm	TWA mg/m³	PEAK/CEIL(C) ppm	PEAK/CEIL(C) mg/m³	AIHA WEELs TWA ppm	TWA mg/m³	STEL/CEIL(C) ppm	STEL/CEIL(C) mg/m³	CARCINOGENICITY CATEGORY
Parathion 56-38-2	0.05 **IFV** Skin; BEI				0.1 Skin				0.05 Skin				0.1 **I** Skin; D		II (8)						EPA-C IARC-3 TLV-A4
Particles (insoluble or poorly soluble) not otherwise specified	*See* Appendix B in *TLVs®* and *BEIs®* book																				
Particulates not otherwise classified/regulated (PNOC; PNOR)					15*, 5** or 50 mppcf* 15 mppcf** *Total dust **Respirable fraction								*See* Dust; general threshold limit value								
Patulin 149-29-1																					IARC-3
Penicillic acid 90-65-3																					IARC-3
Pentaborane 19624-22-7	0.005	0.013	0.015	0.039	0.005	0.01			0.005	0.01	0.015	0.03	0.005	0.013	II (2)						
2,2′,4,4′,5-Pentabromo-diphenyl ether (BDE-99) 60348-60-9																					EPA-II
Pentabromodiphenyl ether 32534-81-9																					EPA-D
Pentachlorobenzene 608-93-5																					EPA-D

SUBSTANCE CAS#	ACGIH® TLVs® TWA ppm	ACGIH® TLVs® TWA mg/m³	ACGIH® TLVs® STEL/CEIL(C) ppm	ACGIH® TLVs® STEL/CEIL(C) mg/m³	OSHA PELs TWA ppm	OSHA PELs TWA mg/m³	OSHA PELs STEL/CEIL(C) ppm	OSHA PELs STEL/CEIL(C) mg/m³	NIOSH RELs TWA ppm	NIOSH RELs TWA mg/m³	NIOSH RELs STEL/CEIL(C) ppm	NIOSH RELs STEL/CEIL(C) mg/m³	DFG MAKs TWA ppm	DFG MAKs TWA mg/m³	DFG MAKs PEAK/CEIL(C) ppm	DFG MAKs PEAK/CEIL(C) mg/m³	AIHA WEELs TWA ppm	AIHA WEELs TWA mg/m³	AIHA WEELs STEL/CEIL(C) ppm	AIHA WEELs STEL/CEIL(C) mg/m³	CARCINOGENICITY CATEGORY
3,4,5,3′,4′-Pentachloro-biphenyl (PCB-126) 57465-28-8																					IARC-1
Pentachlorocyclo-pentadiene 25329-35-5																					EPA-D
2,3,4,7,8-Pentachloro-dibenzofuran 57117-31-4																					IARC-1
Pentachloroethane 76-01-7									handle with caution *See* Pocket Guide App. C				5	42	II (2)						IARC-3
Pentachloronaphthalene 1321-64-8		0.5 Skin				0.5 Skin				0.5 Skin				Skin							
Pentachloronitro-benzene 82-68-8		0.5																			IARC-3 TLV-A4
Pentachlorophenol 87-86-5		0.5 **IFV** Skin; BEI		1 **IFV**		0.5 Skin				0.5 Skin				Skin							EPA-L NTP-R* IARC-2B TLV-A3 MAK-2 *includes by-products of its synthesis
Pentaerythritol 115-77-5		10				15*; 5** *Total dust **Respirable fraction				10*; 5** *Total dust **Respirable fraction											
Pentaerythritol triacrylate 3524-68-3											Sh						1		DSEN		

SUBSTANCE / CAS#	ACGIH® TLVs® TWA ppm	mg/m³	STEL/CEIL(C) ppm	mg/m³	OSHA PELs TWA ppm	mg/m³	STEL/CEIL(C) ppm	mg/m³	NIOSH RELs TWA ppm	mg/m³	STEL/CEIL(C) ppm	mg/m³	DFG MAKs TWA ppm	mg/m³	PEAK/CEIL(C) ppm	mg/m³	AIHA WEELs TWA ppm	mg/m³	STEL/CEIL(C) ppm	mg/m³	CARCINOGENICITY CATEGORY
1,1,1,2,2-Pentafluoro-ethane 354-33-6																	1000				
1,1,1,3,3-Pentafluoro-propane 460-73-1																	300				
Pentane, all isomers 78-78-4; 109-66-0; 463-82-1	1000	2950			1000	2950			120	350	C 610* CAS: 109-66-0 only	C 1800* *15-min	1000	3000	II (2) C						
2,3-Pentanedione 600-14-6													0.02	0.083	II (1) Skin; Sh; D						
2,4-Pentanedione 123-54-6	25	102											20	83	II (2) Skin; C						
		Skin																			
Pentanol, all isomers 71-41-0; 75-84-3; 75-85-4; 123-51-3; 137-32-6; 584-02-1; 598-75-4; 6032-29-7; 30899-19-5; 94624-12-1													20	73	I (2) C		100				
																	CAS: 71-41-0 only				
1-Pentyl acetate (n-Amyl acetate) 628-63-7	50	266	100	532	100	525			100	525			50	270	I (1) C						
2-Pentyl acetate (sec-Amyl acetate) 626-38-0	50	266	100	532	125	650			125	650			50	270	I (1) D						

SUBSTANCE / CAS#	ACGIH® TLVs® TWA ppm	mg/m³	STEL/CEIL(C) ppm	mg/m³	OSHA PELs TWA ppm	mg/m³	STEL/CEIL(C) ppm	mg/m³	NIOSH RELs TWA ppm	mg/m³	STEL/CEIL(C) ppm	mg/m³	DFG MAKs TWA ppm	mg/m³	PEAK/CEIL(C) ppm	mg/m³	AIHA WEELs TWA ppm	mg/m³	STEL/CEIL(C) ppm	mg/m³	CARCINOGENICITY CATEGORY
3-Pentyl acetate 620-11-1	50	266	100	532									50	270	I (1) D						
4-Pentyl acetate (tert-Amyl-acetate) 625-16-1	50	226	100	532									50	270	I (1) D						
Pentyl mercaptan 110-66-7											C 0.5*	C 2.1* *15-min									
Pepsin 9001-75-6														Sa							
Peracetic acid 79-21-0			0.4 IFV	1.24 IFV																	MAK-3B TLV-A4
Perchlorate and perchlorate salts 7601-89-0; 7778-74-7; 7790-98-9; 7791-03-9																					EPA-NL
Perchloromethyl mercaptan 594-42-3	0.1	0.76			0.1	0.8			0.1	0.8											
Perchloryl fluoride 7616-94-6	3	13	6	25	3	13.5			3	14	6	28									
Perfluorobutyl ethylene (PFBE) 19430-93-4	100	1023																			

SUBSTANCE / CAS#	ACGIH® TLVs® TWA ppm	TWA mg/m³	STEL/CEIL(C) ppm	STEL/CEIL(C) mg/m³	OSHA PELs TWA ppm	TWA mg/m³	STEL/CEIL(C) ppm	STEL/CEIL(C) mg/m³	NIOSH RELs TWA ppm	TWA mg/m³	STEL/CEIL(C) ppm	STEL/CEIL(C) mg/m³	DFG MAKs TWA ppm	TWA mg/m³	PEAK/CEIL(C) ppm	PEAK/CEIL(C) mg/m³	AIHA WEELs TWA ppm	TWA mg/m³	STEL/CEIL(C) ppm	STEL/CEIL(C) mg/m³	CARCINOGENICITY CATEGORY
Perfluoroisobutylene 382-21-8			C 0.01	C 0.082																	
Perfluorooctanesulfonic acid (PFOS) and its salts 1763-23-1														0.01 **I**	II (8)						MAK-3B
													Skin; B								
Perfluorooctanoic acid and its inorganic salts 335-67-1														0.005 **I**	II (8)						IARC-2B MAK-4
													Skin; B								
Perlite 93763-70-3		TLV® withdrawn due to insufficient data				15*; 5** *Total dust **Respirable fraction				10*; 5** *Total dust **Respirable fraction											
Permethrin 52645-53-1																					IARC-3
Persulfates, as persulfate 7727-21-1; 7727-27-1		0.1																			
Perylene 198-55-0																					IARC-3
Petasitenine 60102-37-6																					IARC-3
Petroleum distillates, hydrotreated light 64742-47-8													50**	5 R* 350** *Aerosol **Vapor C	II (4)* II (2)**						MAK-3B* *vapor and aerosol

SUBSTANCE / CAS#	ACGIH® TLVs® TWA ppm	mg/m³	STEL/CEIL(C) ppm	mg/m³	OSHA PELs TWA ppm	mg/m³	STEL/CEIL(C) ppm	mg/m³	NIOSH RELs TWA ppm	mg/m³	STEL/CEIL(C) ppm	mg/m³	DFG MAKs TWA ppm	mg/m³	PEAK/CEIL(C) ppm	mg/m³	AIHA WEELs TWA ppm	mg/m³	STEL/CEIL(C) ppm	mg/m³	CARCINOGENICITY CATEGORY
Petroleum distillates, Naphtha (Rubber solvent) 8002-05-9	Rubber solvent TLV® withdrawn				500	2000				350	C 1800* *15-min										IARC-3
Petroleum sulfonates, calcium salts (technical mixture in Minera oil) 61789-86-4													5 R		II (4) D						
Phenacetin 62-44-2																					IARC-1 NTP-R
Phenanthrene 85-01-8													Skin								EPA-D IARC-3
Phenazopyridine hydrochloride 136-40-3																					IARC-2B NTP-R
Phenelzine sulfate 156-51-4																					IARC-3
Phenicarbazide 103-03-7																					IARC-3
Phenol 108-95-2	5	19		Skin; BEI	5	19		Skin	5	19	C 15.6* *15-min Skin	C 60*			Skin; 3B						EPA-I; D IARC-3 MAK-3B TLV-A4
Phenolphthalein 77-09-8																					IARC-2B NTP-R

SUBSTANCE CAS#	ACGIH® TLVs® TWA ppm	TWA mg/m³	STEL/CEIL(C) ppm	STEL/CEIL(C) mg/m³	OSHA PELs TWA ppm	TWA mg/m³	STEL/CEIL(C) ppm	STEL/CEIL(C) mg/m³	NIOSH RELs TWA ppm	TWA mg/m³	STEL/CEIL(C) ppm	STEL/CEIL(C) mg/m³	DFG MAKs TWA ppm	TWA mg/m³	PEAK/CEIL(C) ppm	PEAK/CEIL(C) mg/m³	AIHA WEELs TWA ppm	TWA mg/m³	STEL/CEIL(C) ppm	STEL/CEIL(C) mg/m³	CARCINOGENICITY CATEGORY
Phenothiazine 92-84-2	5								5												
	Skin								Skin												
Phenoxybenzamine hydrochloride 63-92-3																					IARC-2B NTP-R
2-Phenoxyethanol (Ethylene glycol mono-phenyl ether) 122-99-6													1	5.7*	I (1)						
													*can also occur as vapor and aerosol		C						
Phenyl arsenic compounds																		Skin			MAK-3B
Phenylbutazone 50-33-9																					IARC-3
Phenylenediamine, m-isomer 108-45-2		0.1																			IARC-3 MAK-3B TLV-A4
														Skin; Sh							
Phenylenediamine, o-isomer 95-54-5		0.1																			MAK-3B TLV-A3
														Sh							
Phenylenediamine, p-isomer 106-50-3		0.1				0.1				0.1				0.1 I		II (2)					IARC-3 MAK-3B TLV-A4
						Skin				Skin				Skin; Sh; C							
2-Phenyl-1-ethanol 60-12-8														Skin							

SUBSTANCE CAS#	ACGIH® TLVs® TWA ppm	mg/m³	STEL/CEIL(C) ppm	mg/m³	OSHA PELs TWA ppm	mg/m³	STEL/CEIL(C) ppm	mg/m³	NIOSH RELs TWA ppm	mg/m³	STEL/CEIL(C) ppm	mg/m³	DFG MAKs TWA ppm	mg/m³	PEAK/CEIL(C) ppm	mg/m³	AIHA WEELs TWA ppm	mg/m³	STEL/CEIL(C) ppm	mg/m³	CARCINOGENICITY CATEGORY
Phenyl ether, vapor 101-84-8	1	7	2	14	1	7			1	7			1	7.1	I (1) C						
Phenyl ether/biphenyl mixture, vapor 8004-13-5					1	7			1	7											
Phenyl glycidyl ether (PGE) 122-60-1	0.1	0.6 Skin; DSEN			10	60					C 1* C 6* *15-min See Pocket Guide App. A			Skin; Sh							IARC-2B MAK-2 NIOSH-Ca TLV-A3
Phenylhydrazine 100-63-0	0.1	0.44 Skin			5	22 Skin					C 0.14* C 0.6* Skin *120-min See Pocket Guide App. A			Skin; Sh							MAK-3B NIOSH-Ca TLV-A3
Phenyl isocyanate 103-71-9	0.005	0.025 Skin; DSEN; RSEN	0.015	0.1										Sah							
Phenyl mercaptan 108-98-5	0.1	0.45 Skin									C 0.1* C 0.5* *15-min										
N-Phenyl-1-naph-thylamine 90-30-2														Sh							
N-Phenyl-β-naph-thylamine 135-88-6		L					See Pocket Guide App. A							Sh							IARC-3 NIOSH-Ca* MAK-3B TLV-A4 *since metabolized to β-Naphthylamine
Phenylphenol, o-isomer 90-43-7													5 I* *can also occur as vapor and aerosol		I (1) C						IARC-3 MAK-4

SUBSTANCE / CAS#	ACGIH® TLVs® TWA ppm	TWA mg/m³	STEL/CEIL(C) ppm	STEL/CEIL(C) mg/m³	OSHA PELs TWA ppm	TWA mg/m³	STEL/CEIL(C) ppm	STEL/CEIL(C) mg/m³	NIOSH RELs TWA ppm	TWA mg/m³	STEL/CEIL(C) ppm	STEL/CEIL(C) mg/m³	DFG MAKs TWA ppm	TWA mg/m³	PEAK/CEIL(C) ppm	PEAK/CEIL(C) mg/m³	AIHA WEELs TWA ppm	TWA mg/m³	STEL/CEIL(C) ppm	STEL/CEIL(C) mg/m³	CARCINOGENICITY CATEGORY
Phenylphosphine 638-21-1			C 0.05	C 0.23							C 0.05	C 0.25									
Phenyltin compounds													0.0004*	0.002 I		II (2)					MAK-4
*can also be found as vapor Skin; C																					
Phenytoin 57-41-0																					IARC-2B NTP-R
Phorate 298-02-2		0.05 IFV								0.05		0.2									TLV-A4
Skin; BEI_A									Skin												
Phosgene (Carbonyl chloride) 75-44-5	0.1	0.4			0.1	0.4			0.1	0.4	C 0.2*	C 0.8*	0.1	0.41		I (2)					EPA-II
*15-min															C						
Phosphine 7803-51-2	(0.3) NIC-0.05 NIC-0.07 NIC-A4	(0.42)	(1) NIC-C 0.15	(1.4) NIC-C 0.21	0.3	0.4			0.3	0.4	1	1	0.1	0.14		II (2)					EPA-D
(C)															C						
2-Phosphono-1,2-4-butanetricarboxylic acid 37971-36-1																		10			
(H)																					
Phosphoric acid 7664-38-2		1		3		1				1		3		2 I		I (2)					
(C)															C						
Phosphorus-32, as phosphate 14596-37-3																					IARC-1

SUBSTANCE / CAS#	ACGIH® TLVs® TWA ppm	TWA mg/m³	STEL/CEIL(C) ppm	mg/m³	OSHA PELs TWA ppm	mg/m³	STEL/CEIL(C) ppm	mg/m³	NIOSH RELs TWA ppm	mg/m³	STEL/CEIL(C) ppm	mg/m³	DFG MAKs TWA ppm	mg/m³	PEAK/CEIL(C) ppm	mg/m³	AIHA WEELs TWA ppm	mg/m³	STEL/CEIL(C) ppm	mg/m³	CARCINOGENICITY CATEGORY
Phosphorus, White 7723-14-0										0.1				0.01 I	II (2) C						EPA-D
Phosphorus, Yellow 12185-10-3	0.02	0.1				0.1				0.1				0.01 I	II (2) C						
Phosphorus oxy-chloride 10025-87-3	0.1	0.63							0.1	0.6	0.5	3	0.02	0.13	I (1) C						
Phosphorus penta-chloride 10026-13-8	0.1	0.85				1				1				1 I	I (1) C						
Phosphorus penta-sulfide 1314-80-3		1		3		1				1		3									
Phosphorus pentoxide 1314-56-3														2 I	I (2) C						
Phosphorus trichloride 7719-12-2	0.2	1.1	0.5	2.8	0.5	3			0.2	1.5	0.5	3	0.1	0.57	I (1) C						
o-Phthalaldehyde 643-79-8	NIC-C 0.0001* NIC-C 0.0006* *(V) NIC-Skin; DSEN; RSEN																				
Phthalic acid, m-isomer (Isophthalate) 121-91-5														5 I	I (2) C		10* 5 R *Total dust			

SUBSTANCE CAS#	ACGIH® TLVs® TWA ppm	ACGIH® TLVs® TWA mg/m³	ACGIH® TLVs® STEL/CEIL(C) ppm	ACGIH® TLVs® STEL/CEIL(C) mg/m³	OSHA PELs TWA ppm	OSHA PELs TWA mg/m³	OSHA PELs STEL/CEIL(C) ppm	OSHA PELs STEL/CEIL(C) mg/m³	NIOSH RELs TWA ppm	NIOSH RELs TWA mg/m³	NIOSH RELs STEL/CEIL(C) ppm	NIOSH RELs STEL/CEIL(C) mg/m³	DFG MAKs TWA ppm	DFG MAKs TWA mg/m³	DFG MAKs PEAK/CEIL(C) ppm	DFG MAKs PEAK/CEIL(C) mg/m³	AIHA WEELs TWA ppm	AIHA WEELs TWA mg/m³	AIHA WEELs STEL/CEIL(C) ppm	AIHA WEELs STEL/CEIL(C) mg/m³	CARCINOGENICITY CATEGORY
Phthalic anhydride 85-44-9	0.0003*	0.002*	0.0009* *IFV Skin; DSEN; RSEN	0.005*	2	12			1	6				Sa							TLV-A4
Phthalodinitrile, m-isomer 626-17-5	5 IFV									5											
Phthalodinitrile, o-isomer 91-15-6	1 IFV																				
Phytases														Sa							
Picene 213-46-7																					IARC-3
Picloram 1918-02-1		10				15*; 5** *Total dust **Respirable fraction															IARC-3 TLV-A4
2-Picoline 109-06-8																	2		5 Skin		
3-Picoline 108-99-6																	2		5 Skin		
4-Picoline 108-89-4																	2		5 Skin		

SUBSTANCE / CAS#	ACGIH® TLVs®				OSHA PELs				NIOSH RELs				DFG MAKs				AIHA WEELs				CARCINOGENICITY CATEGORY
	TWA		STEL/CEIL(C)		TWA		STEL/CEIL(C)		TWA		STEL/CEIL(C)		TWA		PEAK/CEIL(C)		TWA		STEL/CEIL(C)		
	ppm	mg/m³	ppm	mg/m³	ppm	mg/m³	ppm	mg/m³	ppm	mg/m³	ppm	mg/m³	ppm	mg/m³	ppm	mg/m³	ppm	mg/m³	ppm	mg/m³	
Picric acid (2,4,6-Trinitrophenol) 88-89-1		0.1				0.1				0.1		0.3									MAK-3B
					Skin				Skin				Skin; Sh								
Picryl chloride 88-88-0														Sh							
Pindone (2-Pivalyl-1,3-indandione) 83-26-1		0.1				0.1				0.1											
Pioglitazone 111025-46-8																					IARC-2A
Piperazine and salts, as Piperazine 110-85-0	0.03 IFV	0.1 IFV												CAS: 110-85-0 only							TLV-A4
		DSEN; RSEN												Sah							
Piperazine dihydro-chloride 142-64-3	TLV® withdrawn; *see* Piperazine and salts, as Piperazine								5												
Piperidine 110-89-4																	1				
																	Skin			
Piperonyl butoxide 51-03-6																					IARC-3
Plaster of Paris (Calcium sulfate hemihydrate) 26499-65-0						15*; 5**				10*; 5**											
					*Total dust **Respirable fraction				*Total dust **Respirable fraction												

SUBSTANCE / CAS#	ACGIH® TLVs® TWA ppm	mg/m³	STEL/CEIL(C) ppm	mg/m³	OSHA PELs TWA ppm	mg/m³	STEL/CEIL(C) ppm	mg/m³	NIOSH RELs TWA ppm	mg/m³	STEL/CEIL(C) ppm	mg/m³	DFG MAKs TWA ppm	mg/m³	PEAK/CEIL(C) ppm	mg/m³	AIHA WEELs TWA ppm	mg/m³	STEL/CEIL(C) ppm	mg/m³	CARCINOGENICITY CATEGORY
Platinum, metal 7440-06-4	1								1												
Platinum, soluble salts, as Pt 7440-06-4	0.002				0.002				0.002							C 0.002					
													Chloroplatinates Sah								
Plutonium 7440-07-5																					IARC-1
Polyacrylic acid 9003-01-4																					IARC-3
Polyalphaolefins														5 R	II (4)						
														C							
Polybrominated biphenyls (PBBs) 59536-65-1																					IARC-2A NTP-R
Polychlorinated biphenyls (PCBs) 1336-36-3																					EPA-B2 IARC-1 NIOSH-Ca NTP-R
Polychlorinated dibenzo-p-dioxins, excluding 2,3,7,8-tetrachlorodibenzo-p-dioxin																					IARC-3

SUBSTANCE / CAS#	ACGIH® TLVs® TWA ppm	TWA mg/m³	STEL/CEIL(C) ppm	STEL/CEIL(C) mg/m³	OSHA PELs TWA ppm	TWA mg/m³	STEL/CEIL(C) ppm	STEL/CEIL(C) mg/m³	NIOSH RELs TWA ppm	TWA mg/m³	STEL/CEIL(C) ppm	STEL/CEIL(C) mg/m³	DFG MAKs TWA ppm	TWA mg/m³	PEAK/CEIL(C) ppm	PEAK/CEIL(C) mg/m³	AIHA WEELs TWA ppm	TWA mg/m³	STEL/CEIL(C) ppm	STEL/CEIL(C) mg/m³	CARCINOGENICITY CATEGORY
Polychlorinated dibenzofurans 136677-10-6																					IARC-3
Polychlorophenols and their sodium salts, mixed exposures																					IARC-2B
Polychloroprene 9010-98-4																					IARC-3
Polyethylene 9002-88-4																					IARC-3
Polyethylene glycol (average molecular weight 200-600)													1000 I		II (8) due to possible mist formation, exposure should be minimized C			10 (H); MW > 200 CAS: 25322-68-3			
Polymethylene poly-phenyl isocyanate (Polymeric MDI) 9016-87-9													0.05 I		I (1) Skin; Sah; C						EPA-CBD; D IARC-3 MAK-4
Polymethyl methacrylate 9011-14-7																					IARC-3
Polypropylene 9003-07-0																					IARC-3
Polypropylene glycol(s) 25322-69-4																		10 (H)			

SUBSTANCE / CAS#	ACGIH® TLVs® TWA ppm	TWA mg/m³	STEL/CEIL(C) ppm	STEL/CEIL(C) mg/m³	OSHA PELs TWA ppm	TWA mg/m³	STEL/CEIL(C) ppm	STEL/CEIL(C) mg/m³	NIOSH RELs TWA ppm	TWA mg/m³	STEL/CEIL(C) ppm	STEL/CEIL(C) mg/m³	DFG MAKs TWA ppm	TWA mg/m³	PEAK/CEIL(C) ppm	PEAK/CEIL(C) mg/m³	AIHA WEELs TWA ppm	TWA mg/m³	STEL/CEIL(C) ppm	STEL/CEIL(C) mg/m³	CARCINOGENICITY CATEGORY
Polystyrene 9003-53-6																					IARC-3
Polytetrafluoro-ethylene 9002-84-0																					IARC-3
Polyurethane foams 9009-54-5																					IARC-3
Polyvinyl acetate 9003-20-7																					IARC-3
Polyvinyl alcohol 9002-89-5																					IARC-3
Polyvinyl chloride (PVC) 9002-86-2		1 R												0.3 R* / 4 I *multiplicated with the material density	II (8) C						IARC-3 MAK-4 TLV-A4
Polyvinyl pyrrolidone 9003-39-8																					IARC-3
Ponceau MX 3761-53-3																					IARC-2B
Ponceau 3R 3564-09-8																					IARC-2B

SUBSTANCE / CAS#	ACGIH® TLVs® TWA ppm	ACGIH® TLVs® TWA mg/m³	ACGIH® TLVs® STEL/CEIL(C) ppm	ACGIH® TLVs® STEL/CEIL(C) mg/m³	OSHA PELs TWA ppm	OSHA PELs TWA mg/m³	OSHA PELs STEL/CEIL(C) ppm	OSHA PELs STEL/CEIL(C) mg/m³	NIOSH RELs TWA ppm	NIOSH RELs TWA mg/m³	NIOSH RELs STEL/CEIL(C) ppm	NIOSH RELs STEL/CEIL(C) mg/m³	DFG MAKs TWA ppm	DFG MAKs TWA mg/m³	DFG MAKs PEAK/CEIL(C) ppm	DFG MAKs PEAK/CEIL(C) mg/m³	AIHA WEELs TWA ppm	AIHA WEELs TWA mg/m³	AIHA WEELs STEL/CEIL(C) ppm	AIHA WEELs STEL/CEIL(C) mg/m³	CARCINOGENICITY CATEGORY
Ponceau SX 4548-53-2																					IARC-3
Portland cement 65997-15-1		1 R E			50 mppcf or 15*; 5**				10*; 5**					Dust*							MAK-3B TLV-A4
Potassium bis(2-hydroxy-ethyl)dithiocarbamate 23746-34-1																					IARC-3
Potassium bromate 7758-01-2																		0.1			IARC-2B
Potassium cyanide, as CN 151-50-8											C 4.7* C 5*			5.0 I		II (1)					
Potassium hydroxide 1310-58-3				C 2								C 2									
Potassium titanates, fibrous dust																					MAK-2
Prazepam 2955-38-6																					IARC-3

Portland cement notes:
*Quartz and chromate fractions must be evaluated as such (valid only for low-chromate cement containing < 2 ppm of Cr(VI). See the Cr(VI) cmpds for cement with a higher Cr(VI) content.)

OSHA PELs: *Total dust **Respirable fraction

NIOSH RELs: *Total dust **Respirable fraction

Potassium cyanide, as CN — ACGIH TLVs: See Hydrogen cyanide and Cyanide salts as CN

Potassium cyanide, as CN — OSHA PELs: See Cyanides

Potassium cyanide, as CN — NIOSH RELs: *10-min applies to other Cyanides as CN except Hydrogen cyanide

Potassium cyanide, as CN — DFG MAKs: Skin; C

SUBSTANCE CAS#	ACGIH® TLVs® TWA ppm	ACGIH® TLVs® TWA mg/m³	ACGIH® TLVs® STEL/CEIL(C) ppm	ACGIH® TLVs® STEL/CEIL(C) mg/m³	OSHA PELs TWA ppm	OSHA PELs TWA mg/m³	OSHA PELs STEL/CEIL(C) ppm	OSHA PELs STEL/CEIL(C) mg/m³	NIOSH RELs TWA ppm	NIOSH RELs TWA mg/m³	NIOSH RELs STEL/CEIL(C) ppm	NIOSH RELs STEL/CEIL(C) mg/m³	DFG MAKs TWA ppm	DFG MAKs TWA mg/m³	DFG MAKs PEAK/CEIL(C) ppm	DFG MAKs PEAK/CEIL(C) mg/m³	AIHA WEELs TWA ppm	AIHA WEELs TWA mg/m³	AIHA WEELs STEL/CEIL(C) ppm	AIHA WEELs STEL/CEIL(C) mg/m³	CARCINOGENICITY CATEGORY
Prednimustine 29069-24-7																					IARC-3
Prednisone 53-03-2																					IARC-3
Primidone 125-33-7																					IARC-2B
Printing inks																					IARC-3
Procarbazine hydrochloride 366-70-1																					IARC-2A NTP-R
Prochloraz 67747-09-5																					EPA-C
Proflavine salts																					IARC-3
Pronetalol hydro-chloride 51-02-5																					IARC-3
Propane 74-98-6	See Appendix F in *TLVs®* and *BEIs®* book (D, EX)				1000	1800			1000	1800			1000	1800	II (4) D						

SUBSTANCE / CAS#	ACGIH® TLVs® TWA ppm	mg/m³	STEL/CEIL(C) ppm	mg/m³	OSHA PELs TWA ppm	mg/m³	STEL/CEIL(C) ppm	mg/m³	NIOSH RELs TWA ppm	mg/m³	STEL/CEIL(C) ppm	mg/m³	DFG MAKs TWA ppm	mg/m³	PEAK/CEIL(C) ppm	mg/m³	AIHA WEELs TWA ppm	mg/m³	STEL/CEIL(C) ppm	mg/m³	CARCINOGENICITY CATEGORY
Propane sultone (1,3-Propane sultone) 1120-71-4		L							*See* Pocket Guide App. A				Skin; 3A								IARC-2B TLV-A3 MAK-1 NIOSH-Ca NTP-R
n-Propanol (n-Propyl alcohol) 71-23-8	100	246			200	500			200	500	250	625									TLV-A4
										Skin											
2-Propanol (Isopropanol; Isopropyl alcohol) 67-63-0	200	492	400	984	400	980			400	980	500	1225	200	500	II (2)						IARC-3 TLV-A4
		BEI												C							
Propargyl alcohol 107-19-7	1	2.3							1	2			2	4.7	I (2)						
		Skin								Skin			Skin; D								
Propargyl bromide 106-96-7																	0.1				
																Skin				
Propham 122-42-9																					IARC-3
β-Propiolactone 57-57-8	0.5	1.5			*See* 29 CFR 1910.1003				*See* Pocket Guide App. A				Skin								IARC-2B OSHA-Ca MAK-2 TLV-A3 NIOSH-Ca NTP-R
Propionaldehyde 123-38-6	20	48															20				EPA-II
Propionic acid 79-09-4	10	30							10	30	15	45	10	31	I (2)						
														C							

SUBSTANCE / CAS#	ACGIH® TLVs® TWA ppm	mg/m³	STEL/CEIL(C) ppm	mg/m³	OSHA PELs TWA ppm	mg/m³	STEL/CEIL(C) ppm	mg/m³	NIOSH RELs TWA ppm	mg/m³	STEL/CEIL(C) ppm	mg/m³	DFG MAKs TWA ppm	mg/m³	PEAK/CEIL(C) ppm	mg/m³	AIHA WEELs TWA ppm	mg/m³	STEL/CEIL(C) ppm	mg/m³	CARCINOGENICITY CATEGORY
Propionitrile 107-12-0									6	14											
Propoxur 114-26-1	0.5 **IFV**		BEI_A							0.5				2 **I**	II (8)						TLV-A3
2-Propoxyethanol (Ethylene glycol mono-n-propyl ether) 2807-30-9													20	86	I (2) Skin; C						
2-Propoxyethyl acetate (Ethylene glycol monopropyl ether acetate) 20706-25-6													20	120	I (2) Skin; C						
n-Propyl acetate 109-60-4	(200)	(835)	(250)	(1040) NIC-withdraw adopted TLVs® and *Documentation; see* Propyl acetate isomers	200	840			200	840	250	1050	100	420	I (2)		D				
Propyl acetate isomers 108-21-4; 109-60-4	NIC-100	NIC-417	NIC-150	NIC-626																	
n-Propyl carbamate 627-12-3																					IARC-3
Propylene 115-07-1	500	860																			IARC-3 TLV-A4
Propylene dichloride (1,2-Dichloropropane) 78-87-5	10	46	DSEN		75	350			*See* Pocket Guide App. A												IARC-3 MAK-3B NIOSH-Ca TLV-A4

SUBSTANCE / CAS#	ACGIH® TLVs® TWA ppm	mg/m³	STEL/CEIL(C) ppm	mg/m³	OSHA PELs TWA ppm	mg/m³	STEL/CEIL(C) ppm	mg/m³	NIOSH RELs TWA ppm	mg/m³	STEL/CEIL(C) ppm	mg/m³	DFG MAKs TWA ppm	mg/m³	PEAK/CEIL(C) ppm	mg/m³	AIHA WEELs TWA ppm	mg/m³	STEL/CEIL(C) ppm	mg/m³	CARCINOGENICITY CATEGORY
Propylene glycol 57-55-6																		10			
Propylene glycol dinitrate (PGDN) 6423-43-4	0.05	0.34							0.05	0.3			0.05	0.34	II (1)						
Skin; BEI$_M$									Skin				Skin								
Propylene oxide (1,2-Epoxypropane) 75-56-9	2	4.8			100	240							2	4.8	I (2)						EPA-B2 NTP-R IARC-2B TLV-A3 MAK-4 NIOSH-Ca
DSEN									*See* Pocket Guide App. A				Sh; C								
Propyleneimine (2-Methylaziridine) 75-55-8	0.2	0.5	0.4	1	2	5			2	5											IARC-2B TLV-A3 MAK-2 NIOSH-Ca NTP-R
Skin					Skin				Skin *See* Pocket Guide App. A				Skin; 3B								
n-Propyl mercaptan 107-03-9											C 0.5*	C 1.6*									
											*15-min										
n-Propyl nitrate 627-13-4	25	107	40	172	25	110			25	105	40	170									
BEI$_M$																					
Propylthiouracil 51-52-5																					IARC-2B NTP-R
Ptaquiloside 87625-62-5																					IARC-3
Pyrene 129-00-0																					EPA-D IARC-3
										Skin											

SUBSTANCE CAS#	ACGIH® TLVs® TWA ppm	ACGIH® TLVs® TWA mg/m³	ACGIH® TLVs® STEL/CEIL(C) ppm	ACGIH® TLVs® STEL/CEIL(C) mg/m³	OSHA PELs TWA ppm	OSHA PELs TWA mg/m³	OSHA PELs STEL/CEIL(C) ppm	OSHA PELs STEL/CEIL(C) mg/m³	NIOSH RELs TWA ppm	NIOSH RELs TWA mg/m³	NIOSH RELs STEL/CEIL(C) ppm	NIOSH RELs STEL/CEIL(C) mg/m³	DFG MAKs TWA ppm	DFG MAKs TWA mg/m³	DFG MAKs PEAK/CEIL(C) ppm	DFG MAKs PEAK/CEIL(C) mg/m³	AIHA WEELs TWA ppm	AIHA WEELs TWA mg/m³	AIHA WEELs STEL/CEIL(C) ppm	AIHA WEELs STEL/CEIL(C) mg/m³	CARCINOGENICITY CATEGORY
Pyrethrum 8003-34-7		5				5				5			does not apply for the constituents of insecticides or synthetic derivatives Sh								TLV-A4
Pyridine 110-86-1	1	3.1			5	15			5	15			Skin								IARC-3 MAK-3B TLV-A3
Pyrido[3,4-c]psoralen 85878-62-2																					IARC-3
Pyrimethamine 58-14-0																					IARC-3
Pyrrolidine 123-75-1													Skin								
Quercetin 117-39-5																					IARC-3
Quinoline 91-22-5																	0.001				EPA-L; B2 Skin
Quinone (p-Benzoquinone) 106-51-4	0.1	0.44			0.1	0.4			0.1	0.4			Sh; 3B								IARC-3 MAK-3B
Reserpine 50-55-5																					IARC-3 NTP-R

SUBSTANCE / CAS#	ACGIH® TLVs®				OSHA PELs				NIOSH RELs				DFG MAKs				AIHA WEELs				CARCINOGENICITY CATEGORY
	TWA		STEL/CEIL(C)		TWA		STEL/CEIL(C)		TWA		STEL/CEIL(C)		TWA		PEAK/CEIL(C)		TWA		STEL/CEIL(C)		
	ppm	mg/m³	ppm	mg/m³	ppm	mg/m³	ppm	mg/m³	ppm	mg/m³	ppm	mg/m³	ppm	mg/m³	ppm	mg/m³	ppm	mg/m³	ppm	mg/m³	
Resorcinol 108-46-3	10	45	20	90					10	45	20	90			Sh						IARC-3 TLV-A4
Retrorsine 480-54-6																					IARC-3
Rhodamine B 81-88-9																					IARC-3
Rhodamine 6G 989-38-8																					IARC-3
Rhodium, elemental 7440-16-6		1				0.1				0.1											MAK-3B TLV-A4
Rhodium, insoluble compounds, as Rh		1				0.1				0.1											MAK-3B* TLV-A4 *inorganic only
Rhodium, soluble compounds, as Rh		0.01				0.001				0.001											MAK-3B* TLV-A4 *inorganic only
Ricinus protein															Sa						
Riddelliine 23246-96-0																					NTP-R

SUBSTANCE / CAS#	ACGIH® TLVs® TWA ppm	mg/m³	STEL/CEIL(C) ppm	mg/m³	OSHA PELs TWA ppm	mg/m³	STEL/CEIL(C) ppm	mg/m³	NIOSH RELs TWA ppm	mg/m³	STEL/CEIL(C) ppm	mg/m³	DFG MAKs TWA ppm	mg/m³	PEAK/CEIL(C) ppm	mg/m³	AIHA WEELs TWA ppm	mg/m³	STEL/CEIL(C) ppm	mg/m³	CARCINOGENICITY CATEGORY
Rifampicin 13292-46-1																					IARC-3
Ripazepam 26308-28-1																					IARC-3
Ronnel 299-84-3	5 IFV BEI_A				15				10												TLV-A4
Rosin core solder thermal decomposition products (colophony) 8050-09-7	DSEN; RSEN; L												Sh								
Rosin core solder, pyrolysis products, as formaldehyde	*See* Rosin core solder thermal decomposition products								0.1 *See* Pocket Guide Apps. A and C												NIOSH-Ca* *in presence of Formaldehyde, Acetaldehyde or Malonaldehyde
Rotenone, commercial 83-79-4	5				5				5				Skin								TLV-A4
Rouge	TLV® withdrawn; *see* Iron oxide				15*; 5** *Total dust **Respirable fraction																
Rubber components													Sh								
Rugulosin 23537-16-8																					IARC-3

SUBSTANCE / CAS#	ACGIH® TLVs® TWA ppm	ACGIH® TLVs® TWA mg/m³	ACGIH® TLVs® STEL/CEIL(C) ppm	ACGIH® TLVs® STEL/CEIL(C) mg/m³	OSHA PELs TWA ppm	OSHA PELs TWA mg/m³	OSHA PELs STEL/CEIL(C) ppm	OSHA PELs STEL/CEIL(C) mg/m³	NIOSH RELs TWA ppm	NIOSH RELs TWA mg/m³	NIOSH RELs STEL/CEIL(C) ppm	NIOSH RELs STEL/CEIL(C) mg/m³	DFG MAKs TWA ppm	DFG MAKs TWA mg/m³	DFG MAKs PEAK/CEIL(C) ppm	DFG MAKs PEAK/CEIL(C) mg/m³	AIHA WEELs TWA ppm	AIHA WEELs TWA mg/m³	AIHA WEELs STEL/CEIL(C) ppm	AIHA WEELs STEL/CEIL(C) mg/m³	CARCINOGENICITY CATEGORY
Saccharated iron oxide 8047-67-4																					IARC-3
Saccharin [81-07-2] and its salts																					IARC-3
Safrole 94-59-7																					IARC-2B NTP-R
Scarlet Red 85-83-6																					IARC-3
Selenious acid 7783-00-8																					EPA-D
Selenium [7782-49-2] compounds, as Se		0.2				0.2				0.2* *except Selenium hexafluoride											EPA-D IARC-3
Selenium [7782-49-2], inorganic compounds, as Se														0.02 I Skin; C	II (8)						MAK-3B
Selenium, metal 7782-49-2		0.2								0.2* *except Selenium hexafluoride				0.02 I Skin; C	II (8)						EPA-D IARC-3 MAK-3B
Selenium hexa-fluoride, as Se 7783-79-1	0.05	0.4			0.05	0.4			0.05												IARC-3

SUBSTANCE	ACGIH® TLVs®				OSHA PELs				NIOSH RELs				DFG MAKs				AIHA WEELs				CARCINOGENICITY CATEGORY
	TWA		STEL/CEIL(C)		TWA		STEL/CEIL(C)		TWA		STEL/CEIL(C)		TWA		PEAK/CEIL(C)		TWA		STEL/CEIL(C)		
CAS#	ppm	mg/m³	ppm	mg/m³	ppm	mg/m³	ppm	mg/m³	ppm	mg/m³	ppm	mg/m³	ppm	mg/m³	ppm	mg/m³	ppm	mg/m³	ppm	mg/m³	
Selenium sulfide 7446-34-6		0.2* *as Se				0.2* *as Se				0.2* *as Se				0.02 I Skin; C		II (8)						EPA-B2 IARC-3 MAK-3B NTP-R
Semicarbazide hydro-chloride 563-41-7																					IARC-3
Seneciphylline 480-81-9																					IARC-3
Senkirkine 2318-18-5																					IARC-3
Sepiolite, fibrous dust e.g., 15501-74-3; 18307-23-8																					IARC-3 MAK-3B
Sesquiterpene lactone															Sh						
Sesone (Sodium-2,4-dichlorophenoxyethyl sulfate) 136-78-7		10				15*; 5** *Total dust **Respirable fraction				10*; 5** *Total dust **Respirable fraction											TLV-A4
Shale oils 68308-34-9																					IARC-1
Shikimic acid 138-59-0																					IARC-3

SUBSTANCE / CAS#	ACGIH® TLVs® TWA ppm	TWA mg/m³	STEL/CEIL(C) ppm	STEL/CEIL(C) mg/m³	OSHA PELs TWA ppm	TWA mg/m³	STEL/CEIL(C) ppm	STEL/CEIL(C) mg/m³	NIOSH RELs TWA ppm	TWA mg/m³	STEL/CEIL(C) ppm	STEL/CEIL(C) mg/m³	DFG MAKs TWA ppm	TWA mg/m³	PEAK/CEIL(C) ppm	PEAK/CEIL(C) mg/m³	AIHA WEELs TWA ppm	TWA mg/m³	STEL/CEIL(C) ppm	STEL/CEIL(C) mg/m³	CARCINOGENICITY CATEGORY
Sidestream smoke (Passive smoking at the workplace i.e., second-hand smoke)																					IARC-1 MAK-1 NTP-K
Silica, amorphous, diatomaceous earth, calcined 68895-54-9														0.3 R C							IARC-3
Silica, amorphous, diatomaceous earth, uncalcined 61790-53-2	TLV® withdrawn due to insufficient data on single substance exposure				20 mppcf or $\dfrac{80\ mg/m^3}{\%\ SiO_2}$				6 *See* Pocket Guide App. C					4 I C							IARC-3
Silica, amorphous, precipitated and gel 112926-00-8	TLV® withdrawn due to insufficient data				20 mppcf or $\dfrac{80\ mg/m^3}{\%\ SiO_2}$				6 *See* Pocket Guide App. C												IARC-3
Silica, amorphous, silica fume 69012-64-2	TLV® withdrawn due to insufficient data																				IARC-3
Silica, amorphous, fused 60676-86-0	TLV® withdrawn due to insufficient data				0.05** $\dfrac{30\ mg/m^{3*;\ ***}}{\%\ SiO_2 + 2}$ $\dfrac{250\ mppcf^{**;\ ***}}{\%\ SiO_2 + 5}$ or $\dfrac{10\ mg/m^{3**;\ ***}}{\%\ SiO_2 + 2}$ *Total dust **Respirable dust ***This standard applies to any operations or sectors for which the Respirable crystalline silica standard, 1910.1053, is stayed or is otherwise not in effect.									0.3 R including CAS: 7699-41-4 C							IARC-3

SUBSTANCE / CAS#	ACGIH® TLVs® TWA ppm	mg/m³	STEL/CEIL(C) ppm	mg/m³	OSHA PELs TWA ppm	mg/m³	STEL/CEIL(C) ppm	mg/m³	NIOSH RELs TWA ppm	mg/m³	STEL/CEIL(C) ppm	mg/m³	DFG MAKs TWA ppm	mg/m³	PEAK/CEIL(C) ppm	mg/m³	AIHA WEELs TWA ppm	mg/m³	STEL/CEIL(C) ppm	mg/m³	CARCINOGENICITY CATEGORY
Silica, crystalline, cristobalite 14464-46-1	0.025 **R**				0.05* *Respirable dust 1/2 the value calculated from the respirable dust formulae for Quartz** **This standard applies to any operations or sectors for which the Respirable crystalline silica standard, 1910.1053, is stayed or is otherwise not in effect.				0.05* *Respirable dust See Pocket Guide App. A												IARC-1 NTP-K* MAK-1* TLV-A2 NIOSH-Ca *respirable
Silica, crystalline, α-quartz 14808-60-7	0.025 **R**				0.05** $\dfrac{30\ \text{mg/m}^{3*;\ ***}}{\%\ SiO_2 + 2}$ $\dfrac{250\ \text{mppcf}^{**;\ ***}}{\%\ SiO_2 + 5}$ or $\dfrac{10\ \text{mg/m}^{3**;\ ***}}{\%\ SiO_2 + 2}$ *Total dust **Respirable dust ***This standard applies to any operations or sectors for which the Respirable crystalline silica standard, 1910.1053, is stayed or is otherwise not in effect.				0.05* *Respirable dust See Pocket Guide App. A												IARC-1 MAK-1* NIOSH-Ca NTP-K* TLV-A2 *respirable
Silica, crystalline, tridymite 15468-32-3	TLV® withdrawn due to insufficient data				0.05* *Respirable dust 1/2 the value calculated from the respirable dust formulae for Quartz** **This standard applies to any operations or sectors for which the Respirable crystalline silica standard, 1910.1053, is stayed or is otherwise not in effect.				0.05* *Respirable dust See Pocket Guide App. A												IARC-1 NIOSH-Ca MAK-1* NTP-K* *respirable

SUBSTANCE CAS#	ACGIH® TLVs® TWA ppm	 TWA mg/m³	 STEL/CEIL(C) ppm	 STEL/CEIL(C) mg/m³	OSHA PELs TWA ppm	 TWA mg/m³	 STEL/CEIL(C) ppm	 STEL/CEIL(C) mg/m³	NIOSH RELs TWA ppm	 TWA mg/m³	 STEL/CEIL(C) ppm	 STEL/CEIL(C) mg/m³	DFG MAKs TWA ppm	 TWA mg/m³	 PEAK/CEIL(C) ppm	 PEAK/CEIL(C) mg/m³	AIHA WEELs TWA ppm	 TWA mg/m³	 STEL/CEIL(C) ppm	 STEL/CEIL(C) mg/m³	CARCINOGENICITY CATEGORY
Silica, crystalline, tripoli 1317-95-9	TLV® withdrawn due to insufficient data and unlikely single substance exposure					0.05** $\frac{30\ mg/m^{3}*;\ ***}{\%\ SiO_2 + 2}$ $\frac{250\ mppcf**;\ ***}{\%\ SiO_2 + 5}$ or $\frac{10\ mg/m^{3}**;\ ***}{\%\ SiO_2 + 2}$ *Total dust **Respirable dust ***This standard applies to any operations or sectors for which the Respirable crystalline silica standard, 1910.1053, is stayed or is otherwise not in effect.				0.05* *Respirable dust *See* Pocket Guide App. A											IARC-1 NIOSH-Ca NTP-K* TLV-A2 *respirable
Silicon 7440-21-3	TLV® withdrawn due to insufficient data					15*; 5** *Total dust **Respirable fraction				10*; 5** *Total dust **Respirable fraction											
Silicon carbide, fibrous dust 409-21-2		0.1 f/cc (F) including whiskers				15*; 5** *Total dust **Respirable fraction				10*; 5** *Total dust **Respirable fraction											MAK-2 TLV-A2
Silicon carbide, nonfibrous 409-21-2		10 **I** 3 **R** E																			
Silicon tetrahydride (Silane) 7803-62-5	5	6.6							5	7											
Silver, metal 7440-22-4		0.1 dust and fume				0.01				0.01				0.1 **I** D	II (8)						EPA-D
Silver, salts, as Ag														0.01 **I** D	I (2)						

SUBSTANCE CAS#	ACGIH® TLVs® TWA ppm	ACGIH® TLVs® TWA mg/m³	ACGIH® TLVs® STEL/CEIL(C) ppm	ACGIH® TLVs® STEL/CEIL(C) mg/m³	OSHA PELs TWA ppm	OSHA PELs TWA mg/m³	OSHA PELs STEL/CEIL(C) ppm	OSHA PELs STEL/CEIL(C) mg/m³	NIOSH RELs TWA ppm	NIOSH RELs TWA mg/m³	NIOSH RELs STEL/CEIL(C) ppm	NIOSH RELs STEL/CEIL(C) mg/m³	DFG MAKs TWA ppm	DFG MAKs TWA mg/m³	DFG MAKs PEAK/CEIL(C) ppm	DFG MAKs PEAK/CEIL(C) mg/m³	AIHA WEELs TWA ppm	AIHA WEELs TWA mg/m³	AIHA WEELs STEL/CEIL(C) ppm	AIHA WEELs STEL/CEIL(C) mg/m³	CARCINOGENICITY CATEGORY
Silver, soluble compounds, as Ag 7440-22-4		0.01				0.01				0.01											
Simazine 122-34-9		0.5 I																			IARC-3 TLV-A3
Soapstone					20 mppcf*				6*; 3**												
	TLV® withdrawn; see Documentation for Talc				*< 1% Crystalline				containing < 1% Quartz *Total dust **Respirable dust												
Sodium aluminum fluoride, as F 15096-52-3						2.5				2.5											
Sodium arsenite 7784-46-5		0.01*				0.01*						C 0.002*									EPA-A NTP-K IARC-1 OSHA-Ca MAK-1 TLV-A1 NIOSH-Ca
	*as As				*as As				as As *15-min See Pocket Guide App. A				as As Skin; 3A							
Sodium azide 26628-22-8			C 0.11*	C 0.29**							C 0.1*	C 0.3**		0.2 I		I (2)					TLV-A4
	*as HN₃ vapor **as NaN₃								*as HN₃ **as NaN₃ Skin				D							
Sodium bisulfite 7631-90-5		5								5											IARC-3 TLV-A4
Sodium chloroacetate 3926-62-3																		2.5* *OARS WEEL			

SUBSTANCE / CAS#	ACGIH® TLVs® TWA ppm	ACGIH® TLVs® TWA mg/m³	ACGIH® TLVs® STEL/CEIL(C) ppm	ACGIH® TLVs® STEL/CEIL(C) mg/m³	OSHA PELs TWA ppm	OSHA PELs TWA mg/m³	OSHA PELs STEL/CEIL(C) ppm	OSHA PELs STEL/CEIL(C) mg/m³	NIOSH RELs TWA ppm	NIOSH RELs TWA mg/m³	NIOSH RELs STEL/CEIL(C) ppm	NIOSH RELs STEL/CEIL(C) mg/m³	DFG MAKs TWA ppm	DFG MAKs TWA mg/m³	DFG MAKs PEAK/CEIL(C) ppm	DFG MAKs PEAK/CEIL(C) mg/m³	AIHA WEELs TWA ppm	AIHA WEELs TWA mg/m³	AIHA WEELs STEL/CEIL(C) ppm	AIHA WEELs STEL/CEIL(C) mg/m³	CARCINOGENICITY CATEGORY
Sodium cyanide, as CN 143-33-9	*See* Hydrogen cyanide and Cyanide salts				*See* Cyanides						C 4.7* *10-min applies to other Cyanides as CN except Hydrogen cyanide	C 5*	3.8 **I** Skin; C		II (1)						
Sodium cyclamate 139-05-9																					IARC-3
Sodium diethyldithio-carbamate 148-18-5													2 **I** Sh; D		II (2)						IARC-3
Sodium fluoroacetate 62-74-8	0.05 Skin				0.05 Skin				0.05 Skin	0.15			0.05 **I** Skin; B		II (4)						
Sodium hydroxide 1310-73-2			C 2		2			C 2													
Sodium hypochlorite 7681-52-9																			2		
Sodium metabisulfite 7681-57-4	5								5												TLV-A4
Sodium persulfate, as S₂O₈ 7775-27-1	0.1												Sah								
Sodium-o-phenyl-phenate 132-27-4													2 **I** C		I (1)						IARC-2B MAK-4

SUBSTANCE / CAS#	ACGIH® TLVs® TWA ppm	TWA mg/m³	STEL/CEIL(C) ppm	STEL/CEIL(C) mg/m³	OSHA PELs TWA ppm	TWA mg/m³	STEL/CEIL(C) ppm	STEL/CEIL(C) mg/m³	NIOSH RELs TWA ppm	TWA mg/m³	STEL/CEIL(C) ppm	STEL/CEIL(C) mg/m³	DFG MAKs TWA ppm	TWA mg/m³	PEAK/CEIL(C) ppm	PEAK/CEIL(C) mg/m³	AIHA WEELs TWA ppm	TWA mg/m³	STEL/CEIL(C) ppm	STEL/CEIL(C) mg/m³	CARCINOGENICITY CATEGORY
Sodium pyrithione 3811-73-2; 15922-78-8														1 I	II (2)						
													Skin; B								
Sodium tetraborate, anhydrous 1330-43-4	See Borate compounds, inorganic									1											
Sodium tetraborate, decahydrate 1303-96-4	See Borate compounds, inorganic									5											
Sodium tetraborate, pentahydrate 12179-04-3	See Borate compounds, inorganic									1				5 I	I (1)						
													C								
Sodium trichloro-acetate 650-51-1														2 I	I (1)						
													Skin; C								
Soya bean constituents														Sa							
Spironolactone 52-01-7																					IARC-3
Starch 9005-25-8		10				15*; 5**				10*; 5**											TLV-A4
					*Total dust **Respirable fraction				*Total dust **Respirable dust												
Stearates		10 I 3 R (J)																			TLV-A4

SUBSTANCE / CAS#	ACGIH® TLVs® TWA ppm	ACGIH® TLVs® TWA mg/m³	ACGIH® TLVs® STEL/CEIL(C) ppm	ACGIH® TLVs® STEL/CEIL(C) mg/m³	OSHA PELs TWA ppm	OSHA PELs TWA mg/m³	OSHA PELs STEL/CEIL(C) ppm	OSHA PELs STEL/CEIL(C) mg/m³	NIOSH RELs TWA ppm	NIOSH RELs TWA mg/m³	NIOSH RELs STEL/CEIL(C) ppm	NIOSH RELs STEL/CEIL(C) mg/m³	DFG MAKs TWA ppm	DFG MAKs TWA mg/m³	DFG MAKs PEAK/CEIL(C) ppm	DFG MAKs PEAK/CEIL(C) mg/m³	AIHA WEELs TWA ppm	AIHA WEELs TWA mg/m³	AIHA WEELs STEL/CEIL(C) ppm	AIHA WEELs STEL/CEIL(C) mg/m³	CARCINOGENICITY CATEGORY
Sterigmatocystin 10048-13-2																					IARC-2B
Stoddard solvent 8052-41-3	100	525			500	2900				350		C 1800* *15-min									
Streptozotocin 18883-66-4																					IARC-2B NTP-R
Strontium chromate, as Cr 7789-06-2		(0.0005) NIC-withdraw adopted TLVs® and *Documentation; see* Chromium and inorganic compounds				0.005 *See* 29 CFR 1910.1026				0.0002* *as Cr(VI) *See* Pocket Guide Apps. A and C				as Cr(VI) Skin; Sh; 2							IARC-1 (TLV-A2) MAK-1 NIOSH-Ca NTP-K
Strychnine 57-24-9		0.15				0.15				0.15											
Styrene, monomer (Phenylethylene; Vinyl benzene) 100-42-5	20	85 BEI	40	170	100		C 200; 600* *5-min peak in any 3 hrs		50	215	100	425	20	86	II (2) C						IARC-2B MAK-5 NTP-R TLV-A4
Styrene-acrylonitrile copolymers 9003-54-7																					IARC-3
Styrene-butadiene copolymers 9003-55-8																					IARC-3
Styrene-7,8-oxide (1,2-Epoxyethylbenzene) 96-09-3	NIC-0.2 NIC-A3	NIC-0.98 NIC-Skin; DSEN																			IARC-2A NTP-R

SUBSTANCE / CAS#	ACGIH® TLVs® TWA ppm	mg/m³	STEL/CEIL(C) ppm	mg/m³	OSHA PELs TWA ppm	mg/m³	STEL/CEIL(C) ppm	mg/m³	NIOSH RELs TWA ppm	mg/m³	STEL/CEIL(C) ppm	mg/m³	DFG MAKs TWA ppm	mg/m³	PEAK/CEIL(C) ppm	mg/m³	AIHA WEELs TWA ppm	mg/m³	STEL/CEIL(C) ppm	mg/m³	CARCINOGENICITY CATEGORY
Subtilisins 1395-21-7; 9014-01-1			C 0.00006	as 100% Crystalline active pure enzyme							0.00006* *60-min			Sa							
Succinic acid 110-15-6										1			2 I		I (2) C						
Succinic anhydride 108-30-5																					IARC-3
Succinonitrile 110-61-2									6	20											
Sucrose 57-50-1		10				15*; 5**				10*; 5**											TLV-A4
					*Total dust **Respirable fraction				*Total dust **Respirable fraction												
Sudan I 842-07-9																					IARC-3
Sudan II 3118-97-6																					IARC-3
Sudan III 85-86-9																					IARC-3
Sudan Brown RR 6416-57-5																					IARC-3

SUBSTANCE / CAS#	ACGIH® TLVs® TWA ppm	TWA mg/m³	STEL/CEIL(C) ppm	STEL/CEIL(C) mg/m³	OSHA PELs TWA ppm	TWA mg/m³	STEL/CEIL(C) ppm	STEL/CEIL(C) mg/m³	NIOSH RELs TWA ppm	TWA mg/m³	STEL/CEIL(C) ppm	STEL/CEIL(C) mg/m³	DFG MAKs TWA ppm	TWA mg/m³	PEAK/CEIL(C) ppm	PEAK/CEIL(C) mg/m³	AIHA WEELs TWA ppm	TWA mg/m³	STEL/CEIL(C) ppm	STEL/CEIL(C) mg/m³	CARCINOGENICITY CATEGORY
Sudan Red 7B 6368-72-5																					IARC-3
Sulfafurazole (Sulfisoxazole) 127-69-5																					IARC-3
Sulfallate 95-06-7																					IARC-2B NTP-R
Sulfamethazine 57-68-1																					IARC-3
Sulfamethoxazole 723-46-6																					IARC-3
Sulfasalazine 599-79-1																					IARC-2B
Sulfites																					IARC-3
Sulfometuron methyl 74222-97-2	5																				TLV-A4
Sulfotepp (TEDP) 3689-24-5	0.1 **IFV** Skin; BEI_A				0.2 Skin				0.2 Skin				0.01*	0.13 **I**	II (2) *can also be found as vapor Skin; C						TLV-A4

SUBSTANCE / CAS#	ACGIH® TLVs® TWA ppm	mg/m³	STEL/CEIL(C) ppm	mg/m³	OSHA PELs TWA ppm	mg/m³	STEL/CEIL(C) ppm	mg/m³	NIOSH RELs TWA ppm	mg/m³	STEL/CEIL(C) ppm	mg/m³	DFG MAKs TWA ppm	mg/m³	PEAK/CEIL(C) ppm	mg/m³	AIHA WEELs TWA ppm	mg/m³	STEL/CEIL(C) ppm	mg/m³	CARCINOGENICITY CATEGORY
Sulfur dioxide 7446-09-5			0.25	0.65	5	13			2	5	5	13	1	2.7	I (1) C 1 C 2.7 C						IARC-3 TLV-A4
Sulfur hexafluoride 2551-62-4	1000	5970			1000	6000			1000	6000			1000	6100	II (8) D						
Sulfuric acid 7664-93-9		0.2 T				1				1				0.1 I	I (1) C 0.2 C						IARC-1* NTP-K* MAK-4 TLV-A2* *refers to Sulfuric acid contained in strong inorganic acid mists
Sulfur monochloride 10025-67-9			C 1	C 5.5	1	6					C 1	C 6									
Sulfur pentafluoride 5714-22-7			C 0.01	C 0.10	0.025	0.25					C 0.01	C 0.1									
Sulfur tetrafluoride 7783-60-0			C 0.1	C 0.44							C 0.1	C 0.4									
Sulfuryl fluoride 2699-79-8	5	21	10	42	5	20			5	20	10	40									
Sulprofos 35400-43-2	0.008* *IFV	0.1* Skin; BEI_A								1											TLV-A4
Sunset Yellow FCF 2783-94-0																					IARC-3

SUBSTANCE / CAS#	ACGIH® TLVs® TWA ppm	mg/m³	STEL/CEIL(C) ppm	mg/m³	OSHA PELs TWA ppm	mg/m³	STEL/CEIL(C) ppm	mg/m³	NIOSH RELs TWA ppm	mg/m³	STEL/CEIL(C) ppm	mg/m³	DFG MAKs TWA ppm	mg/m³	PEAK/CEIL(C) ppm	mg/m³	AIHA WEELs TWA ppm	mg/m³	STEL/CEIL(C) ppm	mg/m³	CARCINOGENICITY CATEGORY
Symphytine 22571-95-5																					IARC-3
Synthetic vitreous fibers, continuous filament glass fibers	5 I	1 f/cc (F)							5*												IARC-3* TLV-A4 *glass filament
									*Total fibrous glass, or 3 f/cc TWA (fib ≤ 3.5 µm diam; ≥ 10 µm length)												
Synthetic vitreous fibers, glass wool fibers		1 f/cc (F)							5*												IARC-3 MAK-2 NTP-R* TLV-A3 *inhalable
									*Total fibrous glass, or 3 f/cc TWA (fib ≤ 3.5 µm diam; ≥ 10 µm length)												
Synthetic vitreous fibers, rock wool fibers		1 f/cc (F)																			IARC-3 MAK-2 TLV-A3
Synthetic vitreous fibers, slag wool fibers		1 f/cc (F)																			IARC-3 MAK-3B TLV-A3
Synthetic vitreous fibers, special purpose glass fibers		1 f/cc (F)																			IARC-2B NTP-R* TLV-A3 *inhalable
Synthetic vitreous fibers, refractory ceramic fibers		0.2 f/cc (F)																			EPA-B2 IARC-2B MAK-2 TLV-A2
2,4,5-T (2,4,5-Trichloro-phenoxyacetic acid) 93-76-5	10				10				10				2 I		II (2)						TLV-A4
													Skin; C								
Talc, containing no asbestos fibers 14807-96-6	2 R	E			20 mppcf* *containing < 1% Quartz				2* and < 1% Quartz *Respirable dust												IARC-3 MAK-3B* TLV-A4 *respirable

SUBSTANCE CAS#	ACGIH® TLVs® TWA ppm	ACGIH® TLVs® TWA mg/m³	ACGIH® TLVs® STEL/CEIL(C) ppm	ACGIH® TLVs® STEL/CEIL(C) mg/m³	OSHA PELs TWA ppm	OSHA PELs TWA mg/m³	OSHA PELs STEL/CEIL(C) ppm	OSHA PELs STEL/CEIL(C) mg/m³	NIOSH RELs TWA ppm	NIOSH RELs TWA mg/m³	NIOSH RELs STEL/CEIL(C) ppm	NIOSH RELs STEL/CEIL(C) mg/m³	DFG MAKs TWA ppm	DFG MAKs TWA mg/m³	DFG MAKs PEAK/CEIL(C) ppm	DFG MAKs PEAK/CEIL(C) mg/m³	AIHA WEELs TWA ppm	AIHA WEELs TWA mg/m³	AIHA WEELs STEL/CEIL(C) ppm	AIHA WEELs STEL/CEIL(C) mg/m³	CARCINOGENICITY CATEGORY
Talc, containing asbestos fibers	Use Asbestos TLV–TWA; (K)				Use Asbestos PEL See 29 CFR 1910.1001				See Asbestos												IARC-1 NTP-K OSHA-Ca TLV-A1
Tall oil, distilled 8002-26-4														Sh* *only applies to Tall oil containing Abietic acid							
Tamoxifen 10540-29-1																					IARC-1 NTP-K
Tannic acid [1401-55-4] and tannins																					IARC-3
Tantalum, metal 7440-25-7	TLV® withdrawn due to insufficient data				5				5		10		4 I								MAK-3A* *respirable fraction
Tantalum oxide, dusts, as Ta 1314-61-0	TLV® withdrawn due to insufficient data				5				5		10										
Tartaric acid 87-69-4													2 I		I (2) C						
Tellurium [13494-80-9] and compounds, as Te	0.1* *except Hydrogen telluride				0.1				0.1* *except Tellurium hexafluoride and Bismuth telluride												
Tellurium hexafluoride, as Te 7783-80-4	0.02	0.2			0.02	0.2			0.02	0.2											

SUBSTANCE / CAS#	ACGIH® TLVs® TWA ppm	TWA mg/m³	STEL/CEIL(C) ppm	STEL/CEIL(C) mg/m³	OSHA PELs TWA ppm	TWA mg/m³	STEL/CEIL(C) ppm	STEL/CEIL(C) mg/m³	NIOSH RELs TWA ppm	TWA mg/m³	STEL/CEIL(C) ppm	STEL/CEIL(C) mg/m³	DFG MAKs TWA ppm	TWA mg/m³	PEAK/CEIL(C) ppm	PEAK/CEIL(C) mg/m³	AIHA WEELs TWA ppm	TWA mg/m³	STEL/CEIL(C) ppm	STEL/CEIL(C) mg/m³	CARCINOGENICITY CATEGORY
Temazepam 846-50-4																					IARC-3
Temephos 3383-96-8	1 IFV Skin; BEI_A				15*; 5** *Total dust **Respirable fraction				10*; 5** *Total dust **Respirable fraction												TLV-A4
Teniposide 29767-20-2																					IARC-2A
Terbufos 1307-79-9	0.01 IFV Skin; BEI_A																				TLV-A4
Terephthalic acid (p-Phthalic acid) 100-21-0	10													5 I	I (2) C						
Terpene polychlorinates 8001-50-1																					IARC-3
Terphenyl, o-, m-, p-isomers 84-15-1; 92-06-8; 92-94-4; 26140-60-3			C 0.53	C 5			C 1	C 9			C 0.5	C 5									
2,2′,4,4′-Tetrabromodiphenyl ether (BDE-47) 5436-43-1																					EPA-II
Tetrabromodiphenyl ether 40088-47-9																					EPA-D

SUBSTANCE / CAS#	ACGIH® TLVs® TWA ppm	mg/m³	STEL/CEIL(C) ppm	mg/m³	OSHA PELs TWA ppm	mg/m³	STEL/CEIL(C) ppm	mg/m³	NIOSH RELs TWA ppm	mg/m³	STEL/CEIL(C) ppm	mg/m³	DFG MAKs TWA ppm	mg/m³	PEAK/CEIL(C) ppm	mg/m³	AIHA WEELs TWA ppm	mg/m³	STEL/CEIL(C) ppm	mg/m³	CARCINOGENICITY CATEGORY
1,1,2,2-Tetrabromo-ethane (Acetylene tetrabromide) 79-27-6	0.1 IFV	1.4 IFV			1	14															
Tetra-n-butyltin compounds													0.004*	0.02 I	I (1)						MAK-4
													Skin**; C								
													*can also be found as vapor **for n-butyltin cmpds whose organic ligands are already designated "Sa" or "Sh", these designations also apply								
2,2',5,5'-Tetrachloro-benzidine 15721-02-5																					IARC-3
Tetrachlorocyclopen-tadiene 695-77-2																					EPA-D
2,3,7,8-Tetrachlorodi-benzo-p-dioxin (TCDD) 1746-01-6									*See* Pocket Guide App. A				$1\cdot10^{-8}$ I		II (8) Skin; C						IARC-1 MAK-4 NIOSH-Ca NTP-K
1,1,1,2-Tetrachloro-2,2-difluoroethane (FC-112a) 76-11-9	100	834			500	4170			500	4170			200	1700	II (2) D						
1,1,2,2-Tetrachloro-1,2-difluoroethane (FC-112) 76-12-0	50	417			500	4170			500	4170			200	1700	II (2) D						
1,1,1,2-Tetrachloro-ethane 630-20-6									handle with caution *See* Pocket Guide App. C												EPA-C IARC-2B

SUBSTANCE / CAS#	ACGIH® TLVs® TWA ppm	mg/m³	STEL/CEIL(C) ppm	mg/m³	OSHA PELs TWA ppm	mg/m³	STEL/CEIL(C) ppm	mg/m³	NIOSH RELs TWA ppm	mg/m³	STEL/CEIL(C) ppm	mg/m³	DFG MAKs TWA ppm	mg/m³	PEAK/CEIL(C) ppm	mg/m³	AIHA WEELs TWA ppm	mg/m³	STEL/CEIL(C) ppm	mg/m³	CARCINOGENICITY CATEGORY
1,1,2,2-Tetrachloro-ethane (Acetylene tetrachloride) 79-34-5	1	6.9			5	35			1	7			1	7	II (2)						EPA-L NIOSH-Ca IARC-2B TLV-A3 MAK-3B
		Skin				Skin			Skin *See* Pocket Guide Apps. A and C				Skin; D								
Tetrachloroethylene (Perchloroethylene) 127-18-4	25	170	100	685	100		C 200; 300*		minimize workplace exposure concentrations				10	69	II (2)						EPA-L NTP-R IARC-2A TLV-A3 MAK-3B NIOSH-Ca
		BEI					*5-min peak in any 3 hrs		*See* Pocket Guide App. A				Skin; C								
Tetrachloronaphthalene 1335-88-2		2				2				2											
							Skin			Skin											
2,3,5,6-Tetrachloro-pyridine 2402-79-1																		5			
Tetrachlorosilane 10026-04-7																			C 1		
Tetrachlorvinphos 22248-79-9																					IARC-3
Tetraethylene glycol diacrylate 17831-71-9																		1			
													Sh				Skin; DSEN			
Tetraethylene glycol dimethacrylate 109-17-1																					
													Sh							
Tetraethylene pentamine 112-57-2																		5			
																	Skin; DSEN			

SUBSTANCE / CAS#	ACGIH® TLVs® TWA ppm	mg/m³	STEL/CEIL(C) ppm	mg/m³	OSHA PELs TWA ppm	mg/m³	STEL/CEIL(C) ppm	mg/m³	NIOSH RELs TWA ppm	mg/m³	STEL/CEIL(C) ppm	mg/m³	DFG MAKs TWA ppm	mg/m³	PEAK/CEIL(C) ppm	mg/m³	AIHA WEELs TWA ppm	mg/m³	STEL/CEIL(C) ppm	mg/m³	CARCINOGENICITY CATEGORY
Tetraethyl lead, as Pb 78-00-2		0.1				0.075				0.075				0.05	II (2)						IARC-3 TLV-A4
	Skin				Skin				Skin				Skin; B								
Tetraethyl pyrophosphate (TEPP) 107-49-3	0.01 **IFV**					0.05				0.05			0.005	0.06	II (2)						
	Skin; BEI_A				Skin				Skin				Skin								
1,1,1,2-Tetrafluoroethane (HFC 134a) 811-97-2													1000	4200	II (8)		1000				
													C								
Tetrafluoroethylene (Tetrafluoroethene) 116-14-3	2	8.2																			IARC-2B MAK-2 NTP-R TLV-A3
2,3,3,3-Tetrafluoro-propene 754-12-1													200	950	II (2)		500				
													C								
1,3,3,3-Tetrafluoro-propylene 1645-83-6; 29118-24-9													1000	4700	II (2)		800				
												CAS: 29118-24-9 only		C			CAS: 1645-83-6 only				
Tetraglycidal-4,4'-meth-ylenedianiline 28768-32-3														Sh							
Tetrahydrofuran 109-99-9	50	147	100	295	200	590			200	590	250	735	50	150	I (2)						EPA-S MAK-4 TLV-A3
	Skin												Skin; C								
Tetrahydrofurfuryl alcohol 97-99-4																	0.5				

SUBSTANCE / CAS#	ACGIH® TLVs® TWA ppm	TWA mg/m³	STEL/CEIL(C) ppm	mg/m³	OSHA PELs TWA ppm	TWA mg/m³	STEL/CEIL(C) ppm	mg/m³	NIOSH RELs TWA ppm	TWA mg/m³	STEL/CEIL(C) ppm	mg/m³	DFG MAKs TWA ppm	TWA mg/m³	PEAK/CEIL(C) ppm	mg/m³	AIHA WEELs TWA ppm	TWA mg/m³	STEL/CEIL(C) ppm	mg/m³	CARCINOGENICITY CATEGORY
Tetrahydrofurfuryl methacrylate 2455-24-5																Sh					
Tetrahydronaphthalene 119-64-2													2	11	I (1)	C					
Tetrahydrothiophene (THT) 110-01-0													50	180	I (1)	C					
Tetrakis(hydroxymethyl)-phosphonium chloride 124-64-1		2														DSEN					IARC-3 TLV-A4
Tetrakis(hydroxymethyl)-phosphonium sulfate 55566-30-8		2														DSEN					IARC-3 TLV-A4
Tetramethyl lead, as Pb 75-74-1		0.15				0.075				0.075				0.05	II (2)	Skin; B					IARC-3
		Skin				Skin				Skin											
Tetramethyl succinonitrile 3333-52-6	0.5	2.8			0.5	3			0.5	3						Skin					
		Skin				Skin				Skin											
Tetramethyltin 594-27-4													0.001	0.005*	II (4)						
													*can also occur as vapor and aerosol Skin; D							
Tetranitromethane 509-14-8	0.005	0.04			1	8			1	8						Skin					IARC-2B MAK-2 NTP-R TLV-A3

SUBSTANCE CAS#	ACGIH® TLVs® TWA ppm	ACGIH® TLVs® TWA mg/m³	ACGIH® TLVs® STEL/CEIL(C) ppm	ACGIH® TLVs® STEL/CEIL(C) mg/m³	OSHA PELs TWA ppm	OSHA PELs TWA mg/m³	OSHA PELs STEL/CEIL(C) ppm	OSHA PELs STEL/CEIL(C) mg/m³	NIOSH RELs TWA ppm	NIOSH RELs TWA mg/m³	NIOSH RELs STEL/CEIL(C) ppm	NIOSH RELs STEL/CEIL(C) mg/m³	DFG MAKs TWA ppm	DFG MAKs TWA mg/m³	DFG MAKs PEAK/CEIL(C) ppm	DFG MAKs PEAK/CEIL(C) mg/m³	AIHA WEELs TWA ppm	AIHA WEELs TWA mg/m³	AIHA WEELs STEL/CEIL(C) ppm	AIHA WEELs STEL/CEIL(C) mg/m³	CARCINOGENICITY CATEGORY
Tetra-n-octyltin compounds, as Sn													0.002*	0.0098 I Skin**; D		II (2)					MAK-4
													*can also be found as vapor **for n-octyltin cmpds whose organic ligands are already designated "Sa" or "Sh", these designations also apply								
Tetrasodium pyro-phosphate 7722-88-5	TLV® withdrawn due to insufficient data									5											
Tetryl 479-45-8	1.5				1.5		Skin		1.5		Skin			Skin; Sh							MAK-3B
Thallium [7440-28-0] and soluble compounds, as Tl	0.02 I Skin				0.1 Skin				0.1 Skin												EPA-II
Theobromine 83-67-0																					IARC-3
Theophylline 58-55-9																					IARC-3
Thiabendazole 148-79-8													20 I	C; 5		II (2)					
Thimerosal (Sodium ethylmercurithiosalicylate) 54-64-8	See Mercury, aryl compounds								See Mercury, aryl compounds					Sh							

SUBSTANCE / CAS#	ACGIH® TLVs® TWA ppm	ACGIH® TLVs® TWA mg/m³	ACGIH® TLVs® STEL/CEIL(C) ppm	ACGIH® TLVs® STEL/CEIL(C) mg/m³	OSHA PELs TWA ppm	OSHA PELs TWA mg/m³	OSHA PELs STEL/CEIL(C) ppm	OSHA PELs STEL/CEIL(C) mg/m³	NIOSH RELs TWA ppm	NIOSH RELs TWA mg/m³	NIOSH RELs STEL/CEIL(C) ppm	NIOSH RELs STEL/CEIL(C) mg/m³	DFG MAKs TWA ppm	DFG MAKs TWA mg/m³	DFG MAKs PEAK/CEIL(C) ppm	DFG MAKs PEAK/CEIL(C) mg/m³	AIHA WEELs TWA ppm	AIHA WEELs TWA mg/m³	AIHA WEELs STEL/CEIL(C) ppm	AIHA WEELs STEL/CEIL(C) mg/m³	CARCINOGENICITY CATEGORY
Thioacetamide 62-55-5																					IARC-2B NTP-R
4,4'-Thiobis(6-tert-butyl-m-cresol) 96-69-5		1 I			15*; 5** *Total dust **Respirable fraction				10*; 5** *Total dust **Respirable fraction												TLV-A4
4,4'-Thiodianiline 139-65-1																					IARC-2B MAK-2 NTP-R
Thioglycolates														2 I	II (2) Skin; Sh; C						
Thioglycolic acid 68-11-1	1	3.8 and salts (Skin) NIC-DSEN							1	4 Skin					Skin; Sh						
Thionyl chloride 7719-09-7		C 0.2								C 1	C 5										
Thiotepa 52-24-4																					IARC-1 NTP-K
Thiouracil 141-90-2																					IARC-2B
Thiourea 62-56-6															Sh; SP						IARC-3 MAK-3B NTP-R

SUBSTANCE / CAS#	ACGIH® TLVs® TWA ppm	ACGIH® TLVs® TWA mg/m³	ACGIH® TLVs® STEL/CEIL(C) ppm	ACGIH® TLVs® STEL/CEIL(C) mg/m³	OSHA PELs TWA ppm	OSHA PELs TWA mg/m³	OSHA PELs STEL/CEIL(C) ppm	OSHA PELs STEL/CEIL(C) mg/m³	NIOSH RELs TWA ppm	NIOSH RELs TWA mg/m³	NIOSH RELs STEL/CEIL(C) ppm	NIOSH RELs STEL/CEIL(C) mg/m³	DFG MAKs TWA ppm	DFG MAKs TWA mg/m³	DFG MAKs PEAK/CEIL(C) ppm	DFG MAKs PEAK/CEIL(C) mg/m³	AIHA WEELs TWA ppm	AIHA WEELs TWA mg/m³	AIHA WEELs STEL/CEIL(C) ppm	AIHA WEELs STEL/CEIL(C) mg/m³	CARCINOGENICITY CATEGORY
Thiram (Tetra-methylthiuram disulfide) 137-26-8		0.05 **IFV** DSEN				5				5				1 **I** Sh; C	II (2)						IARC-3 TLV-A4
Tin, metal 7440-31-5		2				2				2											
Tin, organic compounds, as Sn 7440-31-5		0.1 Skin		0.2		0.1				0.1* *except Cyhexatin Skin				0.1 **I** Skin; D	II (2)						TLV-A4
Tin, oxide and inorganic compounds, except SnH₄, as Sn		2				2* *inorganic compounds except oxides				2											
Tin oxides, as Sn 18282-10-5; 21651-19-4		2								2											
Titanium dioxide 13463-67-7		10 Withdrawn from NIC				15* *Total dust				*See* Pocket Guide App. A											IARC-2B MAK-3A NIOSH-Ca TLV-A4
Titanium tetrachloride 7550-45-0																		0.5			
Tobacco, smokeless																					IARC-1 NTP-K
Tolidine, o-isomer (3,3'-Dimethylbenzidine) 119-93-7		Skin								Skin *See* Pocket Guide Apps. A and C	C 0.02* *60-min										IARC-2B TLV-A3 MAK-2 NIOSH-Ca NTP-R

SUBSTANCE / CAS#	ACGIH® TLVs® TWA ppm	mg/m³	STEL/CEIL(C) ppm	mg/m³	OSHA PELs TWA ppm	mg/m³	STEL/CEIL(C) ppm	mg/m³	NIOSH RELs TWA ppm	mg/m³	STEL/CEIL(C) ppm	mg/m³	DFG MAKs TWA ppm	mg/m³	PEAK/CEIL(C) ppm	mg/m³	AIHA WEELs TWA ppm	mg/m³	STEL/CEIL(C) ppm	mg/m³	CARCINOGENICITY CATEGORY
o-Tolidine-based dyes									minimize exposure; handle with caution *See* Pocket Guide App. C												NIOSH-Ca
Toluene (Toluol) 108-88-3	20	75 BEI			200		C 300; 500* *10-min peak per 8-hr shift		100	375	150	560	50	190 Skin; C	II (4)						EPA-II IARC-3 TLV-A4
Toluene-2,4- [584-84-9] or 2,6-diisocyanate [91-08-7] (or as a mixture)	0.001*	0.007* *IFV Skin; DSEN; RSEN; BEI	0.005*	0.035*			C 0.02 CAS: 584-84-9 only	C 0.14	CAS: 584-84-9 only *See* Pocket Guide App. A				can also occur as vapor and aerosol Sah								IARC-2B* NTP-R* MAK-3A TLV-A3 NIOSH-Ca *CAS: 26471-62-5
Toluenesulfonyl chloride, p-isomer 98-59-9																			C 5		
Toluidine hydrochloride, o-isomer 636-21-5																					NTP-R
Toluidine, m-isomer 108-44-1	2	8.8 Skin; BEI_M																			TLV-A4
Toluidine, o-isomer 95-53-4	2	8.8 Skin; BEI_M			5	22 Skin			Skin *See* Pocket Guide App. A				Skin; 3A								IARC-1 NTP-K MAK-1 TLV-A3 NIOSH-Ca
Toluidine, p-isomer 106-49-0	2	8.8 Skin; BEI_M							*See* Pocket Guide App. A				Skin; Sh								MAK-3B NIOSH-Ca TLV-A3
Toremifene 89778-26-7																					IARC-3

SUBSTANCE / CAS#	ACGIH® TLVs® TWA ppm	mg/m³	STEL/CEIL(C) ppm	mg/m³	OSHA PELs TWA ppm	mg/m³	STEL/CEIL(C) ppm	mg/m³	NIOSH RELs TWA ppm	mg/m³	STEL/CEIL(C) ppm	mg/m³	DFG MAKs TWA ppm	mg/m³	PEAK/CEIL(C) ppm	mg/m³	AIHA WEELs TWA ppm	mg/m³	STEL/CEIL(C) ppm	mg/m³	CARCINOGENICITY CATEGORY
Treosulfan 299-75-2																					IARC-1
Triamterene 396-01-0																					IARC-2B
Tribromochloromethane 594-15-0																					EPA-D
Tribromodiphenyl ether 49690-94-0																					EPA-D
Tributyl phosphate 126-73-8	5 IFV BEI_A				5				0.2	2.5			1	11	II (2) Skin; C						MAK-4 TLV-A3
Tri-n-butyltin compounds, as Sn													0.004*	0.02 **I** Skin**; B	**I** (1)						MAK-4
													*can also be found as vapor **for n-butyltin cmpds whose organic ligands are already designated "Sa" or "Sh", these designations also apply								
Tributyltin oxide 56-35-9															*See* Tri-n-butyltin compounds						EPA-D; CBD
Trichlormethine (Trimustine hydrochloride) 817-09-4																					IARC-2B

SUBSTANCE / CAS#	ACGIH® TLVs® TWA ppm	TWA mg/m³	STEL/CEIL(C) ppm	STEL/CEIL(C) mg/m³	OSHA PELs TWA ppm	TWA mg/m³	STEL/CEIL(C) ppm	STEL/CEIL(C) mg/m³	NIOSH RELs TWA ppm	TWA mg/m³	STEL/CEIL(C) ppm	STEL/CEIL(C) mg/m³	DFG MAKs TWA ppm	TWA mg/m³	PEAK/CEIL(C) ppm	PEAK/CEIL(C) mg/m³	AIHA WEELs TWA ppm	TWA mg/m³	STEL/CEIL(C) ppm	STEL/CEIL(C) mg/m³	CARCINOGENICITY CATEGORY
Trichloroacetic acid 76-03-9	0.5	3.34							1	7			0.2	1.4*	I (1)						EPA-S IARC-2B TLV-A3
													*can also occur as vapor and aerosol C								
Trichloroacetonitrile 545-06-2																					IARC-3
1,2,3-Trichlorobenzene 87-61-6													5	38	II (2)						
													Skin; C								
1,2,4-Trichlorobenzene 120-82-1			C 5	C 37							C 5	C 40									EPA-D MAK-3B
													Skin								
1,3,5-Trichlorobenzene 108-70-3													5	38	II (2)						
													Skin; C								
2,3,4-Trichloro-1-butene 2431-50-7																					MAK-2
													Skin								
Trichlorocyclopenta-diene 77323-84-3																					EPA-D
1,1,2-Trichloroethane 79-00-5	10	55			10	45			10	45			10	55	II (2)						EPA-C TLV-A3 IARC-3 MAK-3B NIOSH-Ca
		Skin				Skin			Skin *See* Pocket Guide Apps. A and C					Skin							
Trichloroethylene 79-01-6	10	54	25	135	100		C 200; 300*														EPA-CaH NTP-K IARC-1 TLV-A2 MAK-1 NIOSH-Ca
		BEI				*5-min peak in any 2 hrs			*See* Pocket Guide Apps. A and C					Skin; 3B							

SUBSTANCE CAS#	ACGIH® TLVs TWA ppm	mg/m³	STEL/CEIL(C) ppm	mg/m³	OSHA PELs TWA ppm	mg/m³	STEL/CEIL(C) ppm	mg/m³	NIOSH RELs TWA ppm	mg/m³	STEL/CEIL(C) ppm	mg/m³	DFG MAKs TWA ppm	mg/m³	PEAK/CEIL(C) ppm	mg/m³	AIHA WEELs TWA ppm	mg/m³	STEL/CEIL(C) ppm	mg/m³	CARCINOGENICITY CATEGORY
Trichlorofluoromethane (Fluorotrichloromethane; FC-11) 75-69-4			C 1000	C 5620	1000	5600					C 1000	C 5600	1000	5700	II (2) C						TLV-A4
Trichloronaphthalene 1321-65-9		5	Skin			5	Skin			5	Skin			Skin							
2,4,6-Trichlorophenol (Trichlorophenol) 88-06-2																					EPA-B2 NTP-R
2-(2,4,5-Trichlorophen-oxy)propionic acid (2,4,5-TP) 93-72-1																					EPA-D
1,2,3-Trichloropropane 96-18-4	0.005	0.03			50	300			10	60	Skin *See* Pocket Guide App. A			Skin							EPA-L NTP-R IARC-2A TLV-A2 MAK-2 NIOSH-Ca
Trichlorosilane 10025-78-2																			C 0.5		
1,1,2-Trichloro-1,2,2-trifluoroethane (CFC-113) 76-13-1	1000	7670	1250	9590	1000	7600			1000	7600	1250	9500	500	3900	II (2) D						TLV-A4
Trichlorphon 52-68-6		1 I	BEI_A																		IARC-3 TLV-A4
Triethanolamine 102-71-6		5												5 I	I (2) C						IARC-3

SUBSTANCE / CAS#	ACGIH® TLVs® TWA ppm	mg/m³	STEL/CEIL(C) ppm	mg/m³	OSHA PELs TWA ppm	mg/m³	STEL/CEIL(C) ppm	mg/m³	NIOSH RELs TWA ppm	mg/m³	STEL/CEIL(C) ppm	mg/m³	DFG MAKs TWA ppm	mg/m³	PEAK/CEIL(C) ppm	mg/m³	AIHA WEELs TWA ppm	mg/m³	STEL/CEIL(C) ppm	mg/m³	CARCINOGENICITY CATEGORY
Triethoxysilane 998-30-1																	0.05				
Triethylamine 121-44-8	0.5	2.07	1	4.14	25	100							1	4.2	I (2)						TLV-A4
		Skin												D							
Triethylene glycol 112-27-6													1000 I		II (2)						
													(because formation of a mist is possible, exposure should be minimized) B							
Triethylene glycol diacrylate 1680-21-3																		1			
													Sh					Skin		
Triethylene glycol diglycidyl ether 1954-28-5																					IARC-3
Triethylene glycol dimethacrylate 109-16-0																					
													Sh							
Triethylene glycol monomethyl ether 112-35-6													50 I		II (2)						
													C							
Triethylene tetramine 112-24-3																		1	6		
													Sh					Skin		
1,3,5-Triethylhexahydro-1,3,5-triazine 7779-27-3																					MAK-3B

SUBSTANCE CAS#	ACGIH® TLVs® TWA ppm	mg/m³	STEL/CEIL(C) ppm	mg/m³	OSHA PELs TWA ppm	mg/m³	STEL/CEIL(C) ppm	mg/m³	NIOSH RELs TWA ppm	mg/m³	STEL/CEIL(C) ppm	mg/m³	DFG MAKs TWA ppm	mg/m³	PEAK/CEIL(C) ppm	mg/m³	AIHA WEELs TWA ppm	mg/m³	STEL/CEIL(C) ppm	mg/m³	CARCINOGENICITY CATEGORY
Triethyl phosphate 78-40-0																		7.45			
Trifluorobromomethane (Bromotrifluoromethane) 75-63-8	1000	6090			1000	6100			1000	6100			1000	6200 C	II (8)						
1,1,1-Trifluoroethane 420-46-2																	1000				
2,2,2-Trifluoroethanol 75-89-8																	0.3				
Trifluralin 1582-09-8																					EPA-C IARC-3
Triglycidyl-p-amino-phenol 5026-74-4														Sh							
1,3,5-Triglycidyl-s-triazinetrione 2451-62-9	0.05													Sah							
Triisobutyl phosphate 126-71-6														Sh							
Trimellitic anhydride 552-30-7	0.0005 **IFV** Skin; DSEN; RSEN		0.002 **IFV**						0.005 handle as extremely toxic substance	0.04			0.04 Sa		I (1)						

SUBSTANCE / CAS#	ACGIH® TLVs® TWA ppm	TWA mg/m³	STEL/CEIL(C) ppm	STEL/CEIL(C) mg/m³	OSHA PELs TWA ppm	TWA mg/m³	STEL/CEIL(C) ppm	STEL/CEIL(C) mg/m³	NIOSH RELs TWA ppm	TWA mg/m³	STEL/CEIL(C) ppm	STEL/CEIL(C) mg/m³	DFG MAKs TWA ppm	TWA mg/m³	PEAK/CEIL(C) ppm	PEAK/CEIL(C) mg/m³	AIHA WEELs TWA ppm	TWA mg/m³	STEL/CEIL(C) ppm	STEL/CEIL(C) mg/m³	CARCINOGENICITY CATEGORY
Trimetacresyl phosphate 563-04-2	NIC-0.0033* *IFV	NIC-0.05*																			
Trimethoxysilane 2487-90-3																	0.05				
Trimethylamine 75-50-3	5	12	15	36					10	24	15	36	2	4.9	I (2) C		1				
4,4′,6-Trimethylangelicin plus ultraviolet A radiation 90370-29-9																					IARC-3
2,4,5-Trimethylaniline 137-17-7														Skin							IARC-3 MAK-2
2,4,6-Trimethylaniline 88-05-1																					IARC-3
Trimethylbenzene, all isomers 95-63-6; 108-67-8; 526-73-8	25* *mixed isomers CAS: 25551-13-7	123*							25	125			20	100 C	II (2)						EPA-II
Trimethylchlorosilane 75-77-4																			C 5		
Trimethylhydroqui-none 700-13-0														Sh							

SUBSTANCE / CAS#	ACGIH® TLVs® TWA ppm	mg/m³	STEL/CEIL(C) ppm	mg/m³	OSHA PELs TWA ppm	mg/m³	STEL/CEIL(C) ppm	mg/m³	NIOSH RELs TWA ppm	mg/m³	STEL/CEIL(C) ppm	mg/m³	DFG MAKs TWA ppm	mg/m³	PEAK/CEIL(C) ppm	mg/m³	AIHA WEELs TWA ppm	mg/m³	STEL/CEIL(C) ppm	mg/m³	CARCINOGENICITY CATEGORY
Trimethylolpropane triacrylate 15625-89-5														Sh				1 Skin			
Trimethylolpropane trimethacrylate 3290-92-4																		1 Skin			
Trimethylpentane, all isomers 29222-48-8													100	470 D	II (2)						
Trimethyl phosphate 512-56-1														Skin; 2							MAK-3B
Trimethyl phosphite 121-45-9	2	10							2	10				Skin							
4,5′,8-Trimethylpsoralen 3902-71-4																					IARC-3
Trimethylquinone 935-92-2														Sh							
Trimethyltin compounds													0.001	0.005* Skin**; D	II (4)						

*can also occur as vapor and aerosol
**for methyltin cmpds whose organic ligands are already designated "Sa" or "Sh", these designations also apply

SUBSTANCE / CAS#	ACGIH® TLVs® TWA ppm	mg/m³	STEL/CEIL(C) ppm	mg/m³	OSHA PELs TWA ppm	mg/m³	STEL/CEIL(C) ppm	mg/m³	NIOSH RELs TWA ppm	mg/m³	STEL/CEIL(C) ppm	mg/m³	DFG MAKs TWA ppm	mg/m³	PEAK/CEIL(C) ppm	mg/m³	AIHA WEELs TWA ppm	mg/m³	STEL/CEIL(C) ppm	mg/m³	CARCINOGENICITY CATEGORY
2,4,7-Trinitrofluorenone 129-79-3																					MAK-3B
2,4,6-Trinitrotoluene (TNT) 118-96-7	0.1				1.5				0.5												EPA-C IARC-3 MAK-2
	Skin; BEI_M				Skin				Skin				Skin; Sh; 3B								
Tri-n-octyltin compounds, as Sn													0.002* 0.0098 I		II (2)						MAK-4
													Skin**; B								
													*can also be found as vapor **for n-octyltin cmpds whose organic ligands are already designated "Sa" or "Sh", these designations also apply								
Triorthocresyl phosphate 78-30-8	0.013*	0.02*			0.1				0.1												
	*IFV Skin; BEI_A								Skin												
Triparacresyl phosphate 78-32-0	NIC-0.0033*	NIC-0.05*																			
	*IFV																				
Triphenyl amine 603-34-9	TLV® withdrawn due to insufficient data								5												
Triphenylene 217-59-4																					IARC-3
Triphenyl phosphate 115-86-6	3				3				3												TLV-A4

SUBSTANCE / CAS#	ACGIH® TLVs® TWA ppm	TWA mg/m³	STEL/CEIL(C) ppm	mg/m³	OSHA PELs TWA ppm	mg/m³	STEL/CEIL(C) ppm	mg/m³	NIOSH RELs TWA ppm	mg/m³	STEL/CEIL(C) ppm	mg/m³	DFG MAKs TWA ppm	mg/m³	PEAK/CEIL(C) ppm	mg/m³	AIHA WEELs TWA ppm	mg/m³	STEL/CEIL(C) ppm	mg/m³	CARCINOGENICITY CATEGORY
Triphenyl phosphate isopropylated 68937-41-7														1 I	II (2) C						
Triphenyl phosphine 603-35-0														5 I	II (2) Sh; C						
Tripropylene glycol diacrylate 42978-66-5														Sh							
Trisodium phosphate 7601-54-9																				5	
Trypan Blue 72-57-1																					IARC-2B
Trypsin [9002-07-7] and Chymotrypsin [9004-07-3]														Sa							
Tungsten [7440-33-7] and compounds, in the absence of Cobalt, as W		3 R																			
Tungsten [7440-33-7] and insoluble compounds, as W	See Tungsten and compounds, in the absence of Cobalt								5		10										
Tungsten, soluble compounds, as W	See Tungsten and compounds, in the absence of Cobalt								1		3										

SUBSTANCE / CAS#	ACGIH® TLVs® TWA ppm	TWA mg/m³	STEL/CEIL(C) ppm	mg/m³	OSHA PELs TWA ppm	mg/m³	STEL/CEIL(C) ppm	mg/m³	NIOSH RELs TWA ppm	mg/m³	STEL/CEIL(C) ppm	mg/m³	DFG MAKs TWA ppm	mg/m³	PEAK/CEIL(C) ppm	mg/m³	AIHA WEELs TWA ppm	mg/m³	STEL/CEIL(C) ppm	mg/m³	CARCINOGENICITY CATEGORY
Tungsten carbide, cemented 11107-01-0; 12718-69-3; 37329-49-0									containing 5–15% Cobalt *See* Pocket Guide App. C												
Turpentine [8006-64-2] **and selected monoterpenes** 80-56-8; 127-91-3; 13466-78-9	20	112 DSEN			100	560 CAS: 8006-64-2 only			100	560 CAS: 8006-64-2 only			5	28	II (2) CAS: 8006-64-2 only Skin; Sh; D						TLV-A4
1-Undecanethiol (Undecyl mercaptan) 5332-52-5											C 0.5* *15-min	C 3.9*									
Uracil mustard 66-75-1																					IARC-2B
Uranium, natural [7440-61-1], **soluble and insoluble compounds, as U**		0.2 BEI		0.6	0.05* 0.25**				0.05* 0.2** *Soluble; **Insoluble *See* Pocket Guide App. A			0.6**	The threshold value of the Commission on Radiological Protection of 20 mSv per year or 400 mSv per working lifetime corresponds to about 25 μg uranium/m³ for poorly soluble uranium cmpds and 250 μg uranium/m³ for soluble cmpds (MMAD of 5 μm). Skin; 3A								MAK-2 NIOSH-Ca TLV-A1
Urea 57-13-6																	10				EPA-II
Urethane (Carbamic acid, ethyl ester; Ethyl carbamate) 51-79-6														Skin; 3A							IARC-2A MAK-2 NTP-R
n-Valeraldehyde 110-62-3	50	176							50	175 *See* Pocket Guide App. C											

*Soluble; **Insoluble (ACGIH/OSHA notes)

SUBSTANCE / CAS#	ACGIH® TLVs® TWA ppm	mg/m³	STEL/CEIL(C) ppm	mg/m³	OSHA PELs TWA ppm	mg/m³	STEL/CEIL(C) ppm	mg/m³	NIOSH RELs TWA ppm	mg/m³	STEL/CEIL(C) ppm	mg/m³	DFG MAKs TWA ppm	mg/m³	PEAK/CEIL(C) ppm	mg/m³	AIHA WEELs TWA ppm	mg/m³	STEL/CEIL(C) ppm	mg/m³	CARCINOGENICITY CATEGORY
Vanadium [7440-62-2] **and inorganic compounds**													Inhalable fraction	2							MAK-2
Vanadium pentoxide, as V 1314-62-1	0.05 I						*Respirable dust, as V₂O₅ C 0.5* **Fume, as V₂O₅	C 0.1**			*15-min, except Vanadium metal and Vanadium carbide C 0.05*			2							IARC-2B MAK-2 TLV-A3
Vanillin 121-33-5																		10			IARC-3
Vat Yellow 4 128-66-5																					IARC-3
Vegetable oil mist		TLV® withdrawn due to insufficient data			15*; 5** *Total dust **Respirable fraction				10*; 5** *Total dust **Respirable fraction												
Vinblastine sulfate 143-67-9																					IARC-3
Vincristine sulfate 2068-78-2																					IARC-3
Vinyl acetate 108-05-4	10	35	15	53							C 4* *15-min	C 15*									IARC-2B MAK-3A TLV-A3
Vinyl bromide 593-60-2	0.5	2.2							*See* Pocket Guide App. A												IARC-2A NIOSH-Ca NTP-R TLV-A2

SUBSTANCE / CAS#	ACGIH® TLVs® TWA ppm	TWA mg/m³	STEL/CEIL(C) ppm	STEL/CEIL(C) mg/m³	OSHA PELs TWA ppm	TWA mg/m³	STEL/CEIL(C) ppm	STEL/CEIL(C) mg/m³	NIOSH RELs TWA ppm	TWA mg/m³	STEL/CEIL(C) ppm	STEL/CEIL(C) mg/m³	DFG MAKs TWA ppm	TWA mg/m³	PEAK/CEIL(C) ppm	PEAK/CEIL(C) mg/m³	AIHA WEELs TWA ppm	TWA mg/m³	STEL/CEIL(C) ppm	STEL/CEIL(C) mg/m³	CARCINOGENICITY CATEGORY
Vinylcarbazole 1484-13-5														Sh							
Vinyl chloride (Chloroethylene) 75-01-4	1	2.6			1		5*														EPA-K; A NTP-K IARC-1 OSHA-Ca MAK-1 TLV-A1 NIOSH-Ca
					*avg. not exceeding any 15 min See 29 CFR 1910.1017				See Pocket Guide App. A												
Vinyl chloride-Vinyl acetate copolymers 9003-22-9																					IARC-3
4-Vinyl cyclohexene 100-40-3	0.1	0.44												Skin			1	4.4			IARC-2B MAK-2 TLV-A3
Vinyl cyclohexene dioxide 106-87-6	0.1	0.57 Skin							10	60 Skin See Pocket Guide App. A				Skin						IARC-2B MAK-2 NIOSH-Ca NTP-R	TLV-A3
Vinyl fluoride 75-02-5	1	1.9							1		C 5										IARC-2A NTP-R TLV-A2
									See 29 CFR 1910.1017												
Vinylidene chloride (1,1-Dichloroethylene) 75-35-4	5	20											2	8	II (2)						EPA-I*; MAK-3B S**; C NIOSH-Ca IARC-3 TLV-A4 *oral route; **inhalation
									See Pocket Guide App. A				C							
Vinylidene chloride-Vinyl chloride copolymers 9011-06-7																					IARC-3
Vinylidene fluoride (1,1-Difluoroethylene) 75-38-7	500	1310							1		C 5										IARC-3 MAK-3B TLV-A4
									See 29 CFR 1910.1017												

SUBSTANCE / CAS#	ACGIH® TLVs® TWA ppm	mg/m³	STEL/CEIL(C) ppm	mg/m³	OSHA PELs TWA ppm	mg/m³	STEL/CEIL(C) ppm	mg/m³	NIOSH RELs TWA ppm	mg/m³	STEL/CEIL(C) ppm	mg/m³	DFG MAKs TWA ppm	mg/m³	PEAK/CEIL(C) ppm	mg/m³	AIHA WEELs TWA ppm	mg/m³	STEL/CEIL(C) ppm	mg/m³	CARCINOGENICITY CATEGORY
N-Vinyl-2-pyrrolidone 88-12-0	0.05	0.23											0.02	0.095	II (2)						IARC-3 MAK-4 TLV-A3
													Skin; C								
Vinyltoluene 25013-15-4	50	242	100	483	100	480			100	480			20	98	I (2)						IARC-3 TLV-A4
													D								
Vinyltrichlorosilane 75-94-5																			C 1		
Vitamin K substances 12001-79-5																					IARC-3
VM & P naphtha 8032-32-4										350		C 1800*									TLV-A3
	TLV® withdrawn; see Appendix H in TLVs® and BEIs® book									*15-min											
Warfarin 81-81-2		0.01 I				0.1				0.1			0.0016*	0.02 I	II (8)						
		Skin											*can also be found as vapor Skin; B							
Welding fumes, not otherwise specified									for Welding fumes See Pocket Guide App. A												IARC-2B NIOSH-Ca
White mineral oil (pharmaceutical) 8042-47-5														5 R	II (4)						
													C								
Wollastonite 13983-17-0																					IARC-3

SUBSTANCE / CAS#	ACGIH® TLVs® TWA ppm	TWA mg/m³	STEL/CEIL(C) ppm	STEL/CEIL(C) mg/m³	OSHA PELs TWA ppm	TWA mg/m³	STEL/CEIL(C) ppm	STEL/CEIL(C) mg/m³	NIOSH RELs TWA ppm	TWA mg/m³	STEL/CEIL(C) ppm	STEL/CEIL(C) mg/m³	DFG MAKs TWA ppm	TWA mg/m³	PEAK/CEIL(C) ppm	PEAK/CEIL(C) mg/m³	AIHA WEELs TWA ppm	TWA mg/m³	STEL/CEIL(C) ppm	STEL/CEIL(C) mg/m³	CARCINOGENICITY CATEGORY
Wood dusts, beech and oak		1 I							*See* Wood dusts, all other species												IARC-1 MAK-1 NTP-K TLV-A1
Wood dusts, birch, mahogany, teak, walnut		1 I							*See* Wood dusts, all other species												IARC-1 MAK-3B NTP-K TLV-A2
Wood dusts, all other species		1 I								1 for Wood dust *See* Pocket Guide App. A											IARC-1 NIOSH-Ca NTP-K TLV-A4
Wood dusts, softwood		1 I							*See* Wood dusts, all other species												IARC-1 MAK-3B NTP-K TLV-A4
Wood dusts, western red cedar		0.5 I DSEN; RSEN							*See* Wood dusts, all other species					includes African white wood Sah							IARC-1 MAK-3B NTP-K TLV-A4
Xylanases 37278-89-0														Sa							
Xylene (Dimethylbenzene), o-, m-, p-isomers 95-47-6; 106-42-3; 108-38-3; 1330-20-7	100	434 BEI	150	651	100	435			100	435	150	655	100	440 Skin; D	II (2)						EPA-I IARC-3 TLV-A4
m-Xylene α,α′-diamine 1477-55-0		Skin	C 0.1							Skin	C 0.1			Sh							
Xylidine, mixed isomers 87-59-2; 95-64-7; 108-69-0; 1300-73-8	0.5 IFV Skin; BEI_M	2.5 IFV			5	25 Skin			2	10 Skin				isomers except the 2,4- and 2,6- isomers Skin							MAK-3A TLV-A3

SUBSTANCE / CAS#	ACGIH® TLVs® TWA ppm	mg/m³	STEL/CEIL(C) ppm	mg/m³	OSHA PELs TWA ppm	mg/m³	STEL/CEIL(C) ppm	mg/m³	NIOSH RELs TWA ppm	mg/m³	STEL/CEIL(C) ppm	mg/m³	DFG MAKs TWA ppm	mg/m³	PEAK/CEIL(C) ppm	mg/m³	AIHA WEELs TWA ppm	mg/m³	STEL/CEIL(C) ppm	mg/m³	CARCINOGENICITY CATEGORY
2,4-Xylidine 95-68-1														Skin							IARC-3 MAK-2
2,5-Xylidine 95-78-3														Skin							IARC-3 MAK-3A
2,6-Xylidine (2,6-Dimethylaniline) 87-62-7														Skin							IARC-2B MAK-2
Yellow AB 85-84-7																					IARC-3
Yellow OB 131-79-3																					IARC-3
Yttrium [7440-65-5] and compounds, as Y		1				1				1											
Zalcitabine 7481-89-2																					IARC-2B
Zectran 315-18-4																					IARC-3
Zeolites, excluding erionite 1318-02-1																					IARC-3

SUBSTANCE / CAS#	ACGIH® TLVs® TWA ppm	TWA mg/m³	STEL/CEIL(C) ppm	STEL/CEIL(C) mg/m³	OSHA PELs TWA ppm	TWA mg/m³	STEL/CEIL(C) ppm	STEL/CEIL(C) mg/m³	NIOSH RELs TWA ppm	TWA mg/m³	STEL/CEIL(C) ppm	STEL/CEIL(C) mg/m³	DFG MAKs TWA ppm	TWA mg/m³	PEAK/CEIL(C) ppm	PEAK/CEIL(C) mg/m³	AIHA WEELs TWA ppm	TWA mg/m³	STEL/CEIL(C) ppm	STEL/CEIL(C) mg/m³	CARCINOGENICITY CATEGORY
Zidovudine (AZT) 30516-87-1																					IARC-2B
Zinc and compounds 7440-66-6														0.1 **R** 2 **I** inorganic compounds *Respirable **Inhalable, excluding Zinc chloride	I (4)* I (2)**						EPA-II; D; I
Zinc beryllium silicate, as Be 39413-47-3 *See* Beryllium and compounds, as Be					0.002		C 0.005 0.025* *30-min peak per 8-hr shift				C 0.0005 *See* Pocket Guide App. A				Sah						EPA-B1; NIOSH-Ca CBD**; K* NTP-K IARC-1 TLV-A1 MAK-1 *inhaled; **ingested
Zinc chloride, fume 7646-85-7	1			2	1				1			2		0.1 **R** 2 **I** *Respirable	I (1)* I (4)** **Inhalable						EPA-II
Zinc chromates, as Cr 11103-86-9; 13530-65-9; 37300-23-5	(0.01) NIC-withdraw adopted TLV® and *Documentation; see* Chromium and inorganic compounds				0.005* *CAS: 13530-65-9 only **as CrO₃, CAS: 11103-86-9 and 37300-23-5		C 0.1**		*See* Chromic acid and chromates				*See* Chromium(VI) inorganic compounds, insoluble								IARC-1 NTP-K (TLV-A1)
Zinc oxide 1314-13-2	2 **R**			10 **R**	15*; 5** *Total dust **Respirable fraction				5 Dust only			C 15		0.1 **R** 2 **I** *Respirable	I (4)* I (2)** **Inhalable						EPA-II; D; I
Zinc oxide, fume 1314-13-2					5				5			10		0.1 **R** 2 **I** *Respirable	I (4)* I (2)** **Inhalable						EPA-II; D; I
Zinc pyrithione 13463-41-7														Skin							

SUBSTANCE / CAS#	ACGIH® TLVs® TWA ppm	mg/m³	STEL/CEIL(C) ppm	mg/m³	OSHA PELs TWA ppm	mg/m³	STEL/CEIL(C) ppm	mg/m³	NIOSH RELs TWA ppm	mg/m³	STEL/CEIL(C) ppm	mg/m³	DFG MAKs TWA ppm	mg/m³	PEAK/CEIL(C) ppm	mg/m³	AIHA WEELs TWA ppm	mg/m³	STEL/CEIL(C) ppm	mg/m³	CARCINOGENICITY CATEGORY
Zinc stearate 557-05-1		10				15*; 5**				10*; 5**				0.1 R / 2 I	I (4)* / I (2)**						EPA-II; D; I TLV-A4
					*Total dust	**Respirable fraction			*Total dust	**Respirable fraction			*Respirable	**Inhalable							
Zineb 12122-67-7																					IARC-3
Ziram 137-30-4														0.01 I	I (2)						IARC-3
													Sh; C							
Zirconium, elemental 7440-67-7		5		10										1 I	I (1)						TLV-A4
													Sah; D							
Zirconium [7440-67-7] compounds, as Zr		5		10		5				5		10									TLV-A4
									except Zirconium tetrachloride											
Zirconium [7440-67-7], insoluble compounds										5		10		1 I	I (1)						
									except Zirconium tetrachloride				Sah; D							
Zirconium [7440-67-7], soluble compounds										5		10									
									except Zirconium tetrachloride				Sah							

50-00-0	Formaldehyde
50-07-7	Mitomycin C
50-18-0	Cyclophosphamide
50-29-3	DDT (Dichlorodiphenyltrichloroethane)
50-32-8	Benzo[a]pyrene
50-33-9	Phenylbutazone
50-41-9	Clomiphene citrate
50-44-2	6-Mercaptopurine
50-53-3	Chlorpromazine
50-55-5	Reserpine
50-76-0	Actinomycin D
50-78-2	Acetylsalicylic acid (Aspirin)
51-02-5	Pronetalol hydrochloride
51-03-6	Piperonyl butoxide
51-18-3	2,4,6-tris(1-Aziridinyl)-s-triazine
51-21-8	5-Fluorouracil
51-52-5	Propylthiouracil
51-75-2	Nitrogen mustard (N-Methyl-bis[2-chloroethyl] amine)
51-79-6	Urethane (Carbamic acid, ethyl ester; Ethyl carbamate)
52-01-7	Spironolactone
52-24-4	Thiotepa
52-46-0	Apholate
52-51-7	2-Bromo-2-nitro-1,3-propanediol
52-68-6	Trichlorphon
53-03-2	Prednisone
53-70-3	Dibenz[a,h]anthracene
53-96-3	2-Acetylaminofluorene (2-AAF)
54-05-7	Chloroquine
54-11-5	Nicotine
54-31-9	Furosemide (Frusemide)
54-64-8	Thimerosal (Sodium ethylmercurithiosalicylate)
54-85-3	Isonicotinic acid hydrazine (Isoniazid)
55-18-5	N-Nitrosodiethylamine (NDEA)

55-38-9	Fenthion
55-63-0	Nitroglycerin (NG)
55-86-7	Nitrogen mustard hydrochloride (Mechlorethamine hydrochloride)
55-98-1	1,4-Butanediol dimethanesulfonate (Busulphan; Myleran®)
56-04-2	Methylthiouracil
56-23-5	Carbon tetrachloride (Tetrachloromethane)
56-25-7	Cantharidin
56-35-9	Tributyltin oxide
56-38-2	Parathion
56-53-1	Diethylstilboestrol
56-55-3	Benz[a]anthracene
56-72-4	Coumaphos
56-75-7	Chloramphenicol
56-81-5	Glycerin
57-06-7	Allyl isothiocyanate
57-12-5	Cyanide, free
57-13-6	Urea
57-14-7	1,1-Dimethylhydrazine
57-24-9	Strychnine
57-39-6	tris(2-Methyl-1-aziridinyl)phosphine oxide
57-41-0	Phenytoin
57-50-1	Sucrose
57-55-6	Propylene glycol
57-57-8	β-Propiolactone
57-68-1	Sulfamethazine
57-74-9	Chlordane
57-88-5	Cholesterol
58-08-2	Caffeine
58-14-0	Pyrimethamine
58-55-9	Theophylline
58-89-9	Lindane (γ-Hexachlorocyclohexane)
58-93-5	Hydrochlorothiazide
59-05-2	Methotrexate

59-50-7   p-Chloro-m-cresol
59-87-0   Nitrofural (Nitrofurazone)
59-89-2   N-Nitrosomorpholine (NMOR)
60-09-3   p-Aminoazobenzene
60-11-7   4-Dimethylaminoazobenzene
60-12-8   2-Phenyl-1-ethanol
60-24-2   Mercaptoethanol
60-29-7   Ethyl ether (Diethyl ether)
60-34-4   Methyl hydrazine (Monomethyl hydrazine)
60-35-5   Acetamide
60-56-0   Methimazole
60-57-1   Dieldrin
61-57-4   Niridazole
61-82-5   Amitrole (3-Amino-1,2,4-triazole)
62-23-7   4-Nitrobenzoic acid
62-44-2   Phenacetin
62-50-0   Ethyl methanesulfonate
62-53-3   Aniline
62-55-5   Thioacetamide
62-56-6   Thiourea
62-73-7   Dichlorvos (DDVP)
62-74-8   Sodium fluoroacetate
62-75-9   N-Nitrosodimethylamine (N,N-Dimethylnitrosoamine)
63-25-2   Carbaryl
63-92-3   Phenoxybenzamine hydrochloride
64-17-5   Ethanol (Ethyl alcohol)
64-18-6   Formic acid
64-19-7   Acetic acid
64-67-5   Diethyl sulfate
65-85-0   Benzoic acid
66-27-3   Methyl methanesulfonate
66-75-1   Uracil mustard
67-20-9   Nitrofurantoin

67-45-8   Furazolidone
67-56-1   Methanol (Methyl alcohol)
67-63-0   2-Propanol (Isopropyl alcohol)
67-64-1   Acetone
67-66-3   Chloroform (Trichloromethane)
67-68-5   Dimethyl sulfoxide
67-72-1   Hexachloroethane
68-11-1   Thioglycolic acid
68-12-2   Dimethylformamide
68-22-4   Norethisterone
68-76-8   tris(Aziridinyl)-p-benzoquinone (Triaziquone)
70-25-7   N-Methyl-N′-nitro-N-nitrosoguanidine (MNNG)
70-30-4   Hexachlorophene
71-23-8   n-Propanol (n-Propyl alcohol)
71-36-3   n-Butanol (n-Butyl alcohol)
71-41-0   1-Pentanol (n-Amyl alcohol) [see Pentanol, all isomers]
71-43-2   Benzene
71-55-6   Methyl chloroform (1,1,1-Trichloroethane)
71-58-9   Medroxyprogesterone acetate
72-20-8   Endrin
72-43-5   Methoxychlor
72-54-8   p,p′-Dichlorodiphenyldichloroethane (DDD)
72-55-9   p,p′-Dichlorodiphenyldichloroethylene (DDE)
72-57-1   Trypan Blue
74-31-7   N,N-Diphenyl-p-phenylenediamine
74-82-8   Methane
74-83-9   Methyl bromide
74-84-0   Ethane
74-85-1   Ethylene
74-86-2   Acetylene
74-87-3   Methyl chloride
74-88-4   Methyl iodide
74-89-5   Methylamine

74-90-8	Hydrogen cyanide
74-93-1	Methyl mercaptan (Methanethiol)
74-96-4	Ethyl bromide (Bromoethane)
74-97-5	Chlorobromomethane (Bromochloromethane)
74-98-6	Propane
74-99-7	Methyl acetylene (Propyne)
75-00-3	Ethyl chloride (Chloroethane)
75-01-4	Vinyl chloride (Chloroethylene)
75-02-5	Vinyl fluoride
75-04-7	Ethylamine
75-05-8	Acetonitrile
75-07-0	Acetaldehyde (Acetic aldehyde)
75-08-1	Ethyl mercaptan (Ethanethiol)
75-09-2	Dichloromethane (Methylene chloride)
75-10-5	Difluoromethane
75-12-7	Formamide
75-15-0	Carbon disulfide
75-18-3	Dimethyl sulfide
75-21-8	Ethylene oxide (EtO)
75-25-2	Bromoform (Tribromomethane)
75-27-4	Bromodichloromethane
75-28-5	Isobutane [see Butane, isomers]
75-29-6	2-Chloropropane
75-31-0	Isopropylamine
75-34-3	1,1-Dichloroethane (Ethylidene chloride)
75-35-4	Vinylidene chloride (1,1-Dichloroethylene)
75-36-5	Acetyl chloride
75-37-6	1,1-Difluoroethane
75-38-7	Vinylidene fluoride (1,1-Difluoroethylene)
75-43-4	Dichlorofluoromethane (FC-21)
75-44-5	Phosgene (Carbonyl chloride)
75-45-6	Chlorodifluoromethane (FC-22)
75-47-8	Iodoform

75-50-3	Trimethylamine
75-52-5	Nitromethane
75-55-8	Propyleneimine (2-Methylaziridine)
75-56-9	Propylene oxide (1,2-Epoxypropane)
75-60-5	Cacodylic acid
75-61-6	Difluorodibromomethane
75-62-7	Bromotrichloromethane
75-63-8	Trifluorobromomethane (Bromotrifluoromethane)
75-65-0	tert-Butanol (tert-Butyl alcohol)
75-68-3	1-Chloro-1,1-difluoroethane (FC-142b)
75-69-4	Trichlorofluoromethane (Fluorotrichloromethane; FC-11)
75-71-8	Dichlorodifluoromethane (FC-12)
75-72-9	Chlorotrifluoromethane (FC-13)
75-74-1	Tetramethyl lead, as Pb
75-77-4	Trimethylchlorosilane
75-78-5	Dimethyldichlorosilane
75-79-6	Methyltrichlorosilane
75-83-2	2,2-Dimethyl butane [see Hexane, isomers, other than n-Hexane]
75-84-3	Neopentyl alcohol [see Pentanol, all isomers]
75-85-4	2-Methyl-2-butanol [see Pentanol, all isomers]
75-86-5	Acetone cyanohydrin
75-87-6	Chloral
75-88-7	2-Chloro-1,1,1-trifluoroethane
75-89-8	2,2,2-Trifluoroethanol
75-91-2	tert-Butyl hydroperoxide
75-94-5	Vinyltrichlorosilane
75-99-0	2,2-Dichloropropionic acid
76-01-7	Pentachloroethane
76-03-9	Trichloroacetic acid
76-06-2	Chloropicrin (Trichloronitromethane)
76-11-9	1,1,1,2-Tetrachloro-2,2-difluoroethane (FC-112a)
76-12-0	1,1,2,2-Tetrachloro-1,2-difluoroethane (FC-112)
76-13-1	1,1,2-Trichloro-1,2,2-trifluoroethane (CFC-113)

76-14-2	Dichlorotetrafluoroethane (1,2-Dichloro-1,1,2,2-tetrafluoroethane)
76-15-3	Chloropentafluoroethane
76-22-2	Camphor, synthetic
76-38-0	Methoxyflurane
76-44-8	Heptachlor
77-09-8	Phenolphthalein
77-47-4	Hexachlorocyclopentadiene
77-73-6	Dicyclopentadiene
77-78-1	Dimethyl sulfate
78-00-2	Tetraethyl lead, as Pb
78-10-4	Ethyl silicate (Silicic acid tetraethyl ester)
78-30-8	Triorthocresyl phosphate
78-32-0	Triparacresyl phosphate
78-34-2	Dioxathion
78-59-1	Isophorone
78-78-4	Isopentane [see Pentane, all isomers]
78-79-5	Isoprene
78-81-9	Isobutylamine
78-82-0	Isobutyronitrile
78-83-1	Isobutanol (Isobutyl alcohol)
78-84-2	Isobutyraldehyde
78-86-4	2-Chlorobutane
78-87-5	Propylene dichloride (1,2-Dichloropropane)
78-89-7	2-Chloro-1-propanol [see 1-Chloro-1-propanol and 2-Chloro-1-propanol]
78-92-2	sec-Butanol (sec-Butyl alcohol)
78-93-3	Methyl ethyl ketone (MEK; 2-Butanone)
78-94-4	Methyl vinyl ketone (3-Buten-2-one)
78-95-5	Chloroacetone
78-98-8	Methylglyoxal
79-00-5	1,1,2-Trichloroethane
79-01-6	Trichloroethylene
79-04-9	Chloroacetyl chloride
79-06-1	Acrylamide
79-07-2	2-Chloroacetamide
79-09-4	Propionic acid
79-10-7	Acrylic acid
79-11-8	Monochloroacetic acid
79-20-9	Methyl acetate
79-21-0	Peracetic acid
79-22-1	Methyl chloroformate (Chloroformic acid methyl ester)
79-24-3	Nitroethane
79-27-6	1,1,2,2-Tetrabromoethane (Acetylene tetrabromide)
79-29-8	2,3-Dimethyl butane [see Hexane, isomers, other than n-Hexane]
79-34-5	1,1,2,2-Tetrachloroethane (Acetylene tetrachloride)
79-38-9	Chlorotrifluoroethylene
79-41-4	Methacrylic acid
79-43-6	Dichloroacetic acid
79-44-7	Dimethylcarbamoyl chloride
79-46-9	2-Nitropropane
80-05-7	Bisphenol A (4,4′-Isopropylidenediphenol; BPA)
80-08-0	Dapsone
80-15-9	Cumene hydroperoxide
80-51-3	p,p′-Oxybis(benzenesulfonyl hydrazide)
80-56-8	α-Pinene [see Turpentine and selected monoterpenes]
80-62-6	Methyl methacrylate (Methacrylic acid methyl ester)
81-07-2	Saccharin and its salts
81-15-2	Musk xylene
81-49-2	1-Amino-2,4-dibromoanthraquinone
81-81-2	Warfarin
81-84-5	1,8-Naphthalic anhydride
81-88-9	Rhodamine B
82-28-0	1-Amino-2-methylanthraquinone
82-68-8	Pentachloronitrobenzene
83-26-1	Pindone (2-Pivalyl-1,3-indandione)
83-32-9	Acenaphthene

83-63-6	Diacetylaminoazotoluene
83-66-9	Musk ambrette
83-67-0	Theobromine
83-79-4	Rotenone, commercial
84-15-1	Terphenyl, o-isomer [*see* Terphenyl, o-, m-, p-isomers]
84-65-1	Anthraquinone
84-66-2	Diethyl phthalate
84-74-2	Dibutyl phthalate
85-00-7	Diquat dibromide [*see* Diquat]
85-01-8	Phenanthrene
85-42-7	Hexahydrophthalic anhydride, all isomers
85-44-9	Phthalic anhydride
85-68-7	Butyl benzyl phthalate
85-83-6	Scarlet Red
85-84-7	Yellow AB
85-86-9	Sudan III
86-30-6	N-Nitrosodiphenylamine
86-50-0	Azinphos-methyl
86-54-4	Hydralazine
86-57-7	1-Nitronaphthalene
86-73-7	Fluorene
86-74-8	Carbazole
86-88-4	ANTU (α-Naphthylthiourea)
87-29-6	Cinnamyl anthranilate
87-59-2	2,3-Xylidine [*see* Xylidine, mixed isomers]
87-61-6	1,2,3-Trichlorobenzene
87-62-7	2,6-Xylidine (2,6-Dimethylaniline)
87-68-3	Hexachlorobutadiene
87-69-4	Tartaric acid
87-86-5	Pentachlorophenol
88-05-1	2,4,6-Trimethylaniline
88-06-2	2,4,6-Trichlorophenol (Trichlorophenol)
88-10-8	Diethylcarbamoyl chloride

88-12-0	N-Vinyl-2-pyrrolidone
88-72-2	Nitrotoluene, o-isomer (2-Nitrotoluene)
88-73-3	Nitrochlorobenzene, o-isomer (2-Chloronitrobenzene)
88-85-7	Dinoseb
88-88-0	Picryl chloride
88-89-1	Picric acid (2,4,6-Trinitrophenol)
89-72-5	o-sec-Butylphenol
89-78-1	DL-Menthol, synthetic
90-04-0	Anisidine, o-isomer
90-12-0	1-Methyl naphthalene
90-30-2	N-Phenyl-1-naphthylamine
90-43-7	Phenylphenol, o-isomer
90-65-3	Penicillic acid
90-94-8	Michler's ketone
91-08-7	Toluene-2,6-diisocyanate
91-15-6	Phthalodinitrile, o-isomer
91-17-8	Decahydronaphthalene
91-20-3	Naphthalene
91-22-5	Quinoline
91-23-6	2-Nitroanisole
91-29-2	4-Nitro-4′-aminodiphenylamine-2-sulfonic acid
91-57-6	2-Methylnaphthalene [*see* 1-Methylnaphthalene and 2-Methylnaphthalene]
91-59-8	β-Naphthylamine (2-Naphthylamine)
91-64-5	Coumarin
91-93-0	3,3′-Dimethoxybenzidine-4,4′-diisocyanate
91-94-1	3,3′-Dichlorobenzidine
91-95-2	3,3′-Diaminobenzidine
92-06-8	Terphenyl, m-isomer [*see* Terphenyl, o-, m-, p-isomers]
92-52-4	Biphenyl (Diphenyl)
92-67-1	4-Aminodiphenyl
92-84-2	Phenothiazine
92-87-5	Benzidine

92-93-3   4-Nitrodiphenyl (4-Nitrobiphenyl)

92-94-4   Terphenyl, p-isomer [see Terphenyl, o-, m-, p-isomers]

93-15-2   Methyleugenol

93-72-1   2-(2,4,5-Trichlorophenoxy)propionic acid (2,4,5-TP)

93-76-5   2,4,5-T (2,4,5-Trichlorophenoxyacetic acid)

94-36-0   Benzoyl peroxide (Dibenzoyl peroxide)

94-37-1   Dipentamethylene-thiuram disulfide

94-58-6   Dihydrosafrole

94-59-7   Safrole

94-75-7   2,4-D (2,4-Dichlorophenoxyacetic acid)

95-06-7   Sulfallate

95-13-6   Indene

95-33-0   N-Cyclohexyl-2-benzothiazolesulfenamide

95-47-6   Xylene, o-isomer (1,2-Dimethylbenzene) [see Xylene
(Dimethylbenzene), o-, m-, p-isomers]

95-48-7   Cresol, o-isomer [see Cresol, all isomers]

95-49-8   Chlorotoluene, o-isomer

95-50-1   Dichlorobenzene, o-isomer (1,2-Dichlorobenzene)

95-51-2   Chloroaniline, o-isomer

95-53-4   Toluidine, o-isomer

95-54-5   Phenylenediamine, o-isomer

95-63-6   1,2,4-Trimethylbenzene [see Trimethyl benzene, all isomers]

95-64-7   3,4-Xylidine [see Xylidine, mixed isomers]

95-68-1   2,4-Xylidine

95-69-2   4-Chloro-o-toluidine

95-70-5   2,5-Diaminotoluene (Toluene-2,5-diamine)

95-76-1   3,4-Dichloroaniline

95-78-3   2,5-Xylidine

95-79-4   5-Chloro-o-toluidine

95-80-7   2,4-Diaminotoluene (Toluene-2,4-diamine)

95-83-0   4-Chloro-o-phenylenediamine

96-05-9   Allyl methacrylate

96-09-3   Styrene-7,8-oxide (1,2-Epoxyethylbenzene)

96-12-8   1,2-Dibromo-3-chloropropane (DBCP)

96-13-9   2,3-Dibromo-1-propanol

96-14-0   3-Methylpentane [see Hexane, isomers, other than n-Hexane]

96-18-4   1,2,3-Trichloropropane

96-20-8   2-Aminobutanol

96-22-0   Diethyl ketone

96-23-1   1,3-Dichloro-2-propanol

96-24-2   3-Monochloro-1,2-propanediol

96-29-7   Methyl ethyl ketoxime (2-Butanone oxime)

96-33-3   Methyl acrylate (Acrylic acid, methyl ester)

96-34-4   Chloroacetic acid, methyl ester (Methyl chloroacetate)

96-37-7   Methylcyclopentane [see Hexane, isomers, other than n-Hexane]

96-45-7   Ethylene thiourea

96-48-0 γ-Butyrolactone

96-69-5   4,4'-Thiobis(6-tert-butyl-m-cresol)

97-00-7   1-Chloro-2,4-dinitrobenzene

97-18-7   Bithionol

97-53-0   Eugenol

97-54-1   Isoeugenol and its isomers

97-56-3   o-Aminoazotoluene

97-63-2   Ethyl methacrylate (Methylacrylic acid, ethyl ester)

97-77-8   Disulfiram

97-88-1   n-Butyl methacrylate

97-90-5   Ethylene glycol dimethacrylate

97-99-4   Tetrahydrofurfuryl alcohol

98-00-0   Furfuryl alcohol

98-01-1   Furfural

98-07-7   Benzotrichloride (Benzyl trichloride)

98-29-3   p-tert-Butylcatechol (4-[1,1-Dimethylethyl]-1,2-benzenediol)

98-51-1   p-tert-Butyltoluene

98-54-4   p-tert-Butylphenol

98-57-7   p-Chlorophenyl methyl sulfone

98-59-9   Toluenesulfonyl chloride, p-isomer

98-73-7	4-tert-Butylbenzoic acid
98-82-8	Cumene
98-83-9	α-Methylstyrene
98-86-2	Acetophenone
98-87-3	Benzal chloride (Benzyl dichloride)
98-88-4	Benzoyl chloride
98-95-3	Nitrobenzene
99-08-1	Nitrotoluene, m-isomer (3-Nitrotoluene)
99-54-7	3,4-Dichloronitrobenzene
99-55-8	5-Nitro-o-toluidine (4-Nitro-2-aminotoluene)
99-56-9	1,2-Diamino-4-nitrobenzene
99-57-0	2-Amino-4-nitrophenol
99-59-2	5-Nitro-o-anisidine
99-65-0	Dinitrobenzene, m-isomer [see Dinitrobenzene, all isomers]
99-80-9	N-Methyl-N,4-dinitrosoaniline
99-96-7	Hydroxybenzoic acid
99-97-8	N,N-Dimethyl-p-toluidine
99-99-0	Nitrotoluene, p-isomer (4-Nitrotoluene)
100-00-5	Nitrochlorobenzene, p-isomer (4-Chloronitrobenzene)
100-01-6	Nitroaniline, p-isomer
100-21-0	Terephthalic acid (p-Phthalic acid)
100-25-4	Dinitrobenzene, p-isomer [see Dinitrobenzene, all isomers]
100-37-8	2-Diethylaminoethanol
100-40-3	4-Vinyl cyclohexene
100-41-4	Ethyl benzene
100-42-5	Styrene, monomer (Phenylethylene; Vinyl benzene)
100-44-7	Benzyl chloride
100-51-6	Benzyl alcohol
100-52-7	Benzaldehyde
100-61-8	N-Methyl aniline (Monomethyl aniline)
100-63-0	Phenylhydrazine
100-74-3	N-Ethylmorpholine
100-75-4	N-Nitrosopiperidine (NPIP)

100-97-0	Hexamethylenetetramine
101-14-4	4,4′-Methylene bis(2-chloroaniline) (MBOCA)
101-21-3	Chloropropham
101-25-7	Dinitrosopentamethylenetetramine
101-54-2	4-Aminodiphenylamine
101-55-3	p-Bromodiphenyl ether
101-61-1	4,4′-Methylene bis(N,N′-dimethyl)aniline (Michler's base)
101-68-8	Methylene bisphenyl isocyanate (MDI; Diphenylmethane-4,4′-diisocyanate)
101-72-4	N-Isopropyl-N′-phenyl-p-phenylenediamine
101-77-9	4,4′-Methylene dianiline (4,4′-Diaminodiphenyl-methane)
101-80-4	bis(4-Aminophenyl)ether (4,4′-Oxydianiline; 4,4′-Diaminodiphenyl ether)
101-83-7	Dicyclohexylamine
101-84-8	Phenyl ether, vapor
101-87-1	N-Cyclohexyl-N′-phenyl-p-phenylenediamine
101-90-6	Diglycidyl resorcinol ether
102-50-1	Cresidine, m-isomer
102-54-5	Dicyclopentadienyl iron, as Fe (Ferrocene)
102-71-6	Triethanolamine
102-77-2	2-(4-Morpholinylmercapto)benzothiazole
102-81-8	2-N-Dibutylaminoethanol
103-03-7	Phenicarbazide
103-09-3	2-Ethylhexyl acetate
103-11-7	2-Ethylhexyl acrylate (Acrylic acid, 2-ethylhexyl ester)
103-23-1	Di(2-ethylhexyl)adipate
103-33-3	Azobenzene
103-71-9	Phenyl isocyanate
103-90-2	Acetaminophen (Paracetamol)
104-12-1	4-Chlorophenyl isocyanate
104-54-1	Cinnamyl alcohol
104-55-2	Cinnamaldehyde
104-76-7	2-Ethylhexanol

104-94-9   Anisidine, p-isomer
105-11-3   Benzoquinone dioxime, p-isomer
105-46-4   sec-Butyl acetate
105-55-5   N,N'-Diethylthiourea
105-60-2   Caprolactam
105-74-8   Lauroyl peroxide
106-24-1   Geraniol
106-35-4   Ethyl butyl ketone (3-Heptanone)
106-42-3   Xylene, p-isomer (1,4-Dimethylbenzene) [see Xylene (Dimethyl
            benzene), o-, m-, p-isomers]
106-44-5   Cresol, p-isomer [see Cresol, all isomers]
106-46-7   Dichlorobenzene, p-isomer (1,4-Dichlorobenzene)
106-47-8   Chloroaniline, p-isomer
106-49-0   Toluidine, p-isomer
106-50-3   Phenylenediamine, p-isomer
106-51-4   Quinone (p-Benzoquinone)
106-87-6   Vinyl cyclohexene dioxide
106-88-7   1,2-Epoxybutane (1,2-Butylene oxide)
106-89-8   Epichlorohydrin (1-Chloro-2,3-epoxypropane)
106-91-2   Glycidyl methacrylate
106-92-3   Allyl glycidyl ether (AGE)
106-93-4   Ethylene dibromide (1,2-Dibromoethane)
106-94-5   1-Bromopropane
106-96-7   Propargyl bromide
106-97-8   Butane (n-Butane) [see Butane, isomers]
106-98-9   1-Butene [see Butenes, all isomers]
106-99-0   1,3-Butadiene
107-01-7   2-Butene [see Butenes, all isomers]
107-02-8   Acrolein
107-03-9   n-Propyl mercaptan
107-05-1   Allyl chloride
107-06-2   Ethylene dichloride (1,2-Dichloroethane)
107-07-3   Ethylene chlorohydrin (2-Chloroethanol)

107-12-0   Propionitrile
107-13-1   Acrylonitrile (Vinyl cyanide)
107-14-2   Chloroacetonitrile
107-15-3   Ethylenediamine (1,2-Diaminoethane)
107-16-4   Glycolonitrile
107-18-6   Allyl alcohol (AA)
107-19-7   Propargyl alcohol
107-20-0   Chloroacetaldehyde
107-21-1   Ethylene glycol
107-22-2   Glyoxal
107-25-5   Methyl vinyl ether
107-30-2   Chloromethyl methyl ether (CMME; Methyl chloromethyl ether)
107-31-3   Methyl formate (Formic acid methyl ester)
107-39-1   Diisobutylene
107-41-5   Hexylene glycol
107-49-3   Tetraethyl pyrophosphate (TEPP)
107-66-4   Dibutyl phosphate
107-75-5   Hydroxycitronellal
107-83-5   2-Methylpentane [see Hexane, isomers, other than n-Hexane]
107-87-9   Methyl propyl ketone (2-Pentanone)
107-98-2   1-Methoxy-2-propanol (Propylene glycol monomethyl ether; PGME)
108-03-2   1-Nitropropane
108-05-4   Vinyl acetate
108-08-7   2,4-Dimethylpentane [see Heptane, all isomers]
108-10-1   Methyl isobutyl ketone (Hexone)
108-11-2   Methyl isobutyl carbinol (Methyl amyl alcohol;
            4-Methyl-2-pentanol)
108-18-9   Diisopropylamine
108-20-3   Isopropyl ether
108-21-4   Isopropyl acetate
108-22-5   Isopropenyl acetate
108-24-7   Acetic anhydride
108-30-5   Succinic anhydride

108-31-6	Maleic anhydride
108-38-3	Xylene, m-isomer (1,3-Dimethylbenzene) [see Xylene (Dimethylbenzene), o-, m-, p-isomers]
108-39-4	Cresol, m-isomer [see Cresol, all isomers]
108-42-9	Chloroaniline, m-isomer
108-44-1	Toluidine, m-isomer
108-45-2	Phenylenediamine, m-isomer
108-46-3	Resorcinol
108-60-1	Bis(2-chloro-1-methylethyl)ether
108-65-6	1-Methoxypropyl-2-acetate (Propylene glycol monomethyl ether acetate)
108-67-8	1,3,5-Trimethylbenzene [see Trimethylbenzene, all isomers]
108-69-0	3,5-Xylidine [see Xylidine, mixed isomers]
108-70-3	1,3,5-Trichlorobenzene
108-78-1	Melamine
108-80-5	Isocyanuric acid
108-83-8	Diisobutyl ketone (2,6-Dimethyl-4-heptanone)
108-84-9	sec-Hexyl acetate
108-86-1	Bromobenzene
108-87-2	Methylcyclohexane
108-88-3	Toluene (Toluol)
108-89-4	4-Picoline
108-90-7	Chlorobenzene (Monochlorobenzene)
108-91-8	Cyclohexylamine
108-93-0	Cyclohexanol
108-94-1	Cyclohexanone
108-95-2	Phenol
108-98-5	Phenyl mercaptan
108-99-6	3-Picoline
109-06-8	2-Picoline
109-16-0	Triethylene glycol dimethacrylate
109-17-1	Tetraethylene glycol dimethacrylate
109-53-5	iso-Butyl vinyl ether

109-59-1	2-Isopropoxyethanol (Ethylene glycol isopropyl ether)
109-60-4	n-Propyl acetate
109-63-7	Boron trifluoride diethyl ether [see Boron trifluoride ethers]
109-66-0	Pentane, all isomers
109-69-3	1-Chlorobutane
109-73-9	n-Butylamine
109-74-0	n-Butyronitrile
109-77-3	Malononitrile
109-79-5	n-Butyl mercaptan (Butanethiol)
109-86-4	2-Methoxyethanol (EGME)
109-87-5	Methylal (Dimethoxymethane)
109-89-7	Diethylamine
109-94-4	Ethyl formate (Formic acid, ethyl ester)
109-99-9	Tetrahydrofuran
110-00-9	Furan
110-01-0	Tetrahydrothiophene (THT)
110-12-3	Methyl isoamyl ketone (Methyl-2-hexanone)
110-15-6	Succinic acid
110-19-0	Isobutyl acetate
110-43-0	Methyl n-amyl ketone (2-Heptanone)
110-49-6	2-Methoxyethyl acetate (EGMEA)
110-54-3	n-Hexane (Hexane)
110-57-6	1,4-Dichlorobutene, trans-isomer
110-61-2	Succinonitrile
110-62-3	n-Valeraldehyde
110-65-6	Butynediol
110-66-7	Pentyl mercaptan
110-80-5	2-Ethoxyethanol (EGEE; Cellosolve)
110-82-7	Cyclohexane
110-83-8	Cyclohexene
110-85-0	Piperazine
110-86-1	Pyridine
110-89-4	Piperidine

110-91-8	Morpholine
111-15-9	2-Ethoxyethyl acetate (EGEEA; Cellosolve acetate)
111-27-3	n-Hexyl alcohol
111-30-8	Glutaraldehyde
111-31-9	n-Hexyl mercaptan (n-Hexanethiol)
111-40-0	Diethylenetriamine
111-42-2	Diethanolamine
111-44-4	Dichloroethyl ether (bis[2-Chloroethyl]ether)
111-46-6	Diethylene glycol
111-65-9	n-Octane [see Octane, all isomers]
111-66-0	1-Octene
111-69-3	Adiponitrile
111-76-2	2-Butoxyethanol (EGBE)
111-84-2	Nonane
111-87-5	1-Octanol
111-88-6	n-Octyl mercaptan
111-90-0	2-(2-Ethoxyethoxy)ethanol (Diethylene glycolmonoethyl ether)
111-91-1	bis(2-Chloroethoxy)methane
111-92-2	Dibutylamine
111-96-6	Diethylene glycol dimethyl ether
112-07-2	2-Butoxyethyl acetate (EGBEA)
112-24-3	Triethylene tetramine
112-27-6	Triethylene glycol
112-30-1	Decyl alcohol
112-34-5	Diethylene glycol monobutyl ether (DGBE)
112-35-6	Triethylene glycol monomethyl ether
112-55-0	Dodecyl mercaptan
112-57-2	Tetraethylene pentamine
112-80-1	Oleic acid
114-07-8	Erythromycin
114-26-1	Propoxur
115-02-6	Azaserine
115-07-1	Propylene
115-10-6	Dimethyl ether
115-11-7	Isobutene [see Butenes, all isomers]
115-28-6	Chlorendic acid
115-29-7	Endosulfan
115-32-2	Dicofol
115-77-5	Pentaerythritol
115-86-6	Triphenyl phosphate
115-90-2	Fensulfothion
115-96-8	tris(2-Chloroethyl) phosphate
116-06-3	Aldicarb
116-14-3	Tetrafluoroethylene (Tetrafluoroethene)
116-15-4	Hexafluoropropylene
117-10-2	Dantron (Chrysazin; 1,8-Dihydroxyanthraquinone)
117-39-5	Quercetin
117-79-3	2-Aminoanthraquinone
117-81-7	Di(2-ethylhexyl)phthalate (DEHP)
118-48-9	N-Carboxyanthranilic anhydride
118-52-5	1,3-Dichloro-5,5-dimethylhydantoin
118-74-1	Hexachlorobenzene (HCB)
118-92-3	Anthranilic acid
118-96-7	2,4,6-Trinitrotoluene (TNT)
119-27-7	2,4-Dinitroanisole
119-34-6	2-Nitro-4-aminophenol (4-Amino-2-nitrophenol)
119-61-9	Benzophenone
119-64-2	Tetrahydronaphthalene
119-90-4	3,3'-Dimethoxybenzidine (o-Dianisidine-based dyes; 3,3'-Dimethoxybenzidine, dyes metabolized to this compound)
119-93-7	Tolidine, o-isomer (3,3'-Dimethylbenzidine)
120-12-7	Anthracene
120-58-1	Isosafrole
120-61-6	Dimethyl terephthalate
120-71-8	Cresidine, p-isomer (5-Methyl-o-anisidine)

120-78-5	2,2'-Dibenzothiazyl disulfide
120-80-9	Catechol (Pyrocatechol)
120-82-1	1,2,4-Trichlorobenzene
120-83-2	2,4-Dichlorophenol
121-14-2	2,4-Dinitrotoluene
121-33-5	Vanillin
121-44-8	Triethylamine
121-45-9	Trimethyl phosphite
121-66-4	2-Amino-5-nitrothiazole
121-69-7	Dimethylaniline (N,N-Dimethylaniline)
121-73-3	Nitrochlorobenzene, m-isomer (1-Chloronitrobenzene)
121-75-5	Malathion
121-82-4	Cyclonite (RDX)
121-88-0	2-Amino-5-nitrophenol
121-91-5	Phthalic acid, m-isomer (Isophthalate)
121-92-6	3-Nitrobenzoic acid
122-34-9	Simazine
122-39-4	Diphenylamine
122-40-7	α-Amylcinnamaldehyde
122-42-9	Propham
122-60-1	Phenyl glycidyl ether (PGE)
122-66-7	Hydrazobenzene (1,2-Diphenylhydrazine)
122-99-6	2-Phenoxyethanol (Ethylene glycol monophenyl ether)
123-09-1	p-Chlorophenyl methyl sulfide
123-19-3	Dipropyl ketone
123-30-8	p-Aminophenol
123-31-9	Hydroquinone (Dihydroxybenzene)
123-33-1	Maleic hydrazide
123-38-6	Propionaldehyde
123-42-2	Diacetone alcohol (4-Hydroxy-4-methyl-2-pentanone)
123-51-3	Isoamyl alcohol
123-54-6	2,4-Pentanedione
123-72-8	Butyraldehyde

123-73-9	trans-Crotonaldehyde [see Crotonaldehyde]
123-75-1	Pyrrolidine
123-86-4	n-Butyl acetate
123-91-1	1,4-Dioxane (Diethylene dioxide)
123-92-2	Isopentyl acetate (Isoamyl acetate)
124-02-7	Diallylamine
124-04-9	Adipic acid
124-09-4	1,6-Hexanediamine
124-17-4	Butyl carbitol acetate (Diethylene glycol monobutyl ether acetate)
124-38-9	Carbon dioxide
124-40-3	Dimethylamine
124-48-1	Chlorodibromomethane
124-58-3	Methylarsonic acid
124-64-1	Tetrakis(hydroxymethyl)phosphonium chloride
124-68-5	2-Amino-2-methyl-1-propanol
125-33-7	Primidone
126-07-8	Griseofulvin
126-71-6	Triisobutyl phosphate
126-73-8	Tributyl phosphate
126-72-7	tris(2,3-Dibromopropyl) phosphate
126-85-2	Nitrogen mustard N-oxide
126-98-7	Methylacrylonitrile
126-99-8	β-Chloroprene (2-Chloro-1,3-butadiene)
127-00-4	1-Chloro-2-propanol
127-07-1	Hydroxyurea
127-18-4	Tetrachloroethylene (Perchloroethylene)
127-19-5	N,N-Dimethylacetamide
127-69-5	Sulfafurazole (Sulfisoxazole)
127-91-3	β-Pinene [see Turpentine and selected monoterpenes]
128-37-0	Butylated hydroxytoluene (BHT; 2,6-Di-tert-butyl-p-cresol)
128-66-5	Vat Yellow 4
129-00-0	Pyrene
129-06-6	Sodium warfarin [see Warfarin]

129-15-7   2-Methyl-1-nitroanthraquinone, uncertain purity
129-16-8   Merbromin
129-17-9   Blue VRS
129-20-4   Oxyphenbutazone
129-43-1   1-Hydroxyanthraquinone
129-79-3   2,4,7-Trinitrofluorenone
131-11-3   Dimethylphthalate
131-79-3   Yellow OB
132-27-4   Sodium-o-phenylphenate
132-32-1   3-Amino-9-ethylcarbazole
132-64-9   Dibenzofuran
132-65-0   Dibenzothiophene
133-06-2   Captan
133-07-3   Folpet
134-29-2   Anisidine hydrochloride, o-isomer
134-32-7   α-Naphthylamine (1-Naphthylamine)
135-20-6   Cupferron
135-88-6   N-Phenyl-β-naphthylamine
136-35-6   Diazaminobenzene
136-40-3   Phenazopyridine hydrochloride
136-78-7   Sesone (Sodium-2,4-dichlorophenoxyethyl sulfate)
137-05-3   Methyl 2-cyanoacrylate
137-17-7   2,4,5-Trimethylaniline
137-26-8   Thiram (Tetramethylthiuram disulfide)
137-30-4   Ziram
137-32-6   2-Methyl-1-butanol [see Pentanol, all isomers]
138-22-7   n-Butyl lactate
138-59-0   Shikimic acid
138-86-3   DL-Limonene
139-05-9   Sodium cyclamate
139-13-9   Nitrilotriacetic acid
139-65-1   4,4′-Thiodianiline
139-94-6   Nithiazide

140-11-4   Benzyl acetate
140-56-7   p-Dimethylaminoazobenzenediazo sodium sulfonate
140-57-8   Aramite®
140-66-9   4-tert-Octylphenol
140-88-5   Ethyl acrylate (Acrylic acid, ethyl ester)
141-32-2   n-Butyl acrylate (Acrylic acid ester, n-Butyl ester)
141-37-7   3,4-Epoxy-6-methylcyclohexylmethyl-3,4-epoxy-6-methyl-cyclohexane carboxylate
141-43-5   Ethanolamine (2-Aminoethanol)
141-66-2   Dicrotophos
141-78-6   Ethyl acetate
141-79-7   Mesityl oxide
141-90-2   Thiouracil
142-64-3   Piperazine dihydrochloride
142-82-5   n-Heptane [see Heptane, all isomers]
142-83-6   2,4-Hexadienal
143-07-7   Lauric acid
143-10-2   Decylmercaptan
143-33-9   Sodium cyanide, as CN
143-50-0   Chlordecone
143-67-9   Vinblastine sulfate
144-34-3   Methyl selenac
144-62-7   Oxalic acid
148-01-6   3,5-Dinitro-o-toluamide (Dinitolmide)
148-18-5   Sodium diethyldithiocarbamate
148-24-3   8-Hydroxyquinoline
148-79-8   Thiabendazole
148-82-3   Melphalan
149-29-1   Patulin
149-30-4   2-Mercaptobenzothiazole
149-57-5   2-Ethylhexanoic acid
150-13-0   p-Aminobenzoic acid
150-68-5   Monuron

150-69-6	Dulcin
150-76-5	4-Methoxyphenol
151-50-8	Potassium cyanide, as CN
151-56-4	Ethyleneimine
151-67-7	Halothane
154-93-8	bis(Chloroethyl)nitrosourea (BCNU)
156-10-5	p-Nitrosodiphenylamine
156-51-4	Phenelzine sulfate
156-59-2	cis-1,2-Dichloroethylene [see 1,2-Dichloroethylene, all isomers]
156-60-5	trans-1,2-Dichloroethylene [see 1,2-Dichloroethylene, all isomers]
156-62-7	Calcium cyanamide
189-55-9	Dibenzo[a,i]pyrene
189-64-0	Dibenzo[a,h]pyrene
191-07-1	Coronene
191-24-2	Benzo[ghi]perylene
191-26-4	Anthanthrene
191-30-0	Dibenzo[a,l]pyrene
192-47-2	Dibenzo[h,rst]pentaphene
192-51-8	Dibenzo[e,l]pyrene
192-65-4	Dibenzo[a,e]pyrene
192-97-2	Benzo[e]pyrene
193-09-9	Naphtho[2,3-e]pyrene
193-39-5	Indeno[1,2,3,cd]pyrene
194-59-2	7H-Dibenzo[c,g]carbazole
195-19-7	Benzo[c]phenanthrene
196-78-1	Benzo[g]chrysene
198-55-0	Perylene
202-33-5	Benz[j]aceanthrylene
202-94-8	11H-Benz[bc]aceanthrylene
202-98-2	4H-Cyclopenta[def]chrysene
203-12-3	Benzo[ghi]fluoranthene
203-20-3	Naphtho[2,1-a]fluoranthene
203-33-8	Benzo[a]fluoranthene
205-12-9	Benzo[c]fluorene
205-82-3	Benzo[j]fluoranthene
205-99-2	Benzo[b]fluoranthene
206-44-0	Fluoranthene
207-08-9	Benzo[k]fluoranthene
207-83-0	13H-Dibenzo[a,g]fluorene
208-96-8	Acenaphthylene
211-91-6	Benz[l]aceanthrylene
213-46-7	Picene
214-17-5	Benzo[b]chrysene
215-58-7	Dibenz[a,c]anthracene
217-59-4	Triphenylene
218-01-9	Chrysene
224-41-9	Dibenz[a,j]anthracene
224-42-0	Dibenz[a,j]acridine
224-53-3	Dibenz[c,h]acridine
225-11-6	Benz[a]acridine
225-51-4	Benz[c]acridine
226-36-8	Dibenz[a,h]acridine
238-84-6	Benzo[a]fluorene
239-35-0	Benzo[b]naphtho[2,1-d]-thiophene
243-17-4	Benzo[b]fluorene
262-12-4	Dibenzo-p-dioxin
271-89-6	Benzofuran
287-92-3	Cyclopentane
298-00-0	Methyl parathion
298-02-2	Phorate
298-04-4	Disulfoton
298-81-7	8-Methoxypsoralen(Methoxsalen) plus ultraviolet A radiation
299-75-2	Treosulfan
299-84-3	Ronnel
299-86-5	Crufomate

300-76-5 Naled (Dibrom; Dimethyl-1,2-dibromo-2,2-dichloroethyl-phosphate)
302-01-2 Hydrazine
302-17-0 Chloral hydrate
303-34-4 Lasiocarpine
303-47-9 Ochratoxin A
305-03-3 Chlorambucil
306-83-2 2,2-Dichloro-1,1,1-trifluoroethane (FC-123)
309-00-2 Aldrin
311-45-5 Diethyl-p-nitrophenyl phosphate
313-67-7 Aristolochic acid
314-13-6 Evans Blue
314-40-9 Bromacil
315-18-4 Zectran
315-22-0 Monocrotaline
319-84-6 α-Hexachlorocyclohexan
319-85-7 β-Hexachlorocyclohexane [see 1,2,3,4,5,6-Hexachlorocyclohexane, mixture of isomers]
319-86-8 Δ-Hexachlorocyclohexane
320-67-2 Azacitidine
330-54-1 Diuron
330-55-2 Linuron
331-39-5 Caffeic acid
333-41-5 Diazinon
334-88-3 Diazomethane
335-67-1 Perfluorooctanoic acid and its inorganic salts
353-42-4 Boron trifluoride dimethyl ether [see Boron trifluoride ethers]
353-50-4 Carbonyl fluoride
354-33-6 1,1,1,2,2-Pentafluoroethane
366-70-1 Procarbazine hydrochloride
373-02-4 Nickel acetate
382-21-8 Perfluoroisobutylene
396-01-0 Triamterene
406-90-6 Fluroxene
409-21-2 Silicon carbide, fibrous and nonfibrous forms (including whiskers)
420-04-2 Cyanamide
420-12-2 Ethylene sulfide
420-46-2 1,1,1-Trifluoroethane
434-07-1 Oxymetholone
439-14-5 Diazepam
443-48-1 Metronidazole
446-86-6 Azathioprine
460-19-5 Cyanogen
460-73-1 1,1,1,3,3-Pentafluoropropane
463-51-4 Ketene
463-58-1 Carbonyl sulfide
463-82-1 tert-Pentane [see Pentane, all isomers]
471-34-1 Calcium carbonate, synthetic [see Calcium carbonate]
479-45-8 Tetryl
480-54-6 Retrorsine
480-81-9 Seneciphylline
484-20-8 5-Methoxypsoralen
492-17-1 2,4'-Diphenyldiamine
492-80-8 Auramine
493-52-7 Methyl Red
494-03-1 N,N-bis(2-Chloroethyl)-2-naphthylamine (Chlornaphazine)
494-38-2 Acridine Orange
501-30-4 Kojic acid
504-24-5 4-Aminopyridine
504-29-0 2-Aminopyridine
505-29-3 1,4-Dithiane
505-60-2 Mustard gas (2,2'-Dichlorodiethyl sulfide)
506-68-3 Cyanogen bromide
506-77-4 Cyanogen chloride
507-20-0 tert-Butylchloride
509-14-8 Tetranitromethane

510-15-6	Chlorobenzilate
512-56-1	Trimethyl phosphate
513-37-1	1-Chloro-2-methylpropene (Dimethylvinyl chloride)
514-10-3	Abietic acid
518-75-2	Citrinin
520-18-3	Kempferol
523-44-4	Orange I
523-50-2	Angelicin plus ultraviolet A radiation
526-73-8	1,2,3-Trimethylbenzene [see Trimethylbenzene, all isomers]
528-29-0	Dinitrobenzene, o-isomer [see Dinitrobenzene, all isomers]
531-76-0	Merphalan
531-82-8	Furothiazole
532-27-4	2-Chloroacetophenone (Phenacyl chloride)
532-82-1	Chrysoidine
534-52-1	4,6-Dinitro-o-cresol
536-33-4	Ethionamide
538-75-0	1,3-Dicyclohexylcarbodiimide
540-59-0	1,2-Dichloroethylene, all isomers (Acetylene dichloride)
540-73-8	1,2-Dimethylhydrazine
540-84-1	Isooctane (2,2,4-Trimethylpentane) [see Octane, all isomers]
540-88-5	tert-Butyl acetate
541-02-6	Decamethylcyclopentasiloxane
541-41-3	Ethyl chloroformate (Chloroformic acid ethyl ester)
541-73-1	Dichlorobenzene, m-isomer
541-85-5	Ethyl amyl ketone (5-Methyl-3-heptanone)
542-56-3	Isobutyl nitrite
542-75-6	1,3-Dichloropropene
542-78-9	Malonaldehyde
542-88-1	bis(Chloromethyl)ether
542-92-7	Cyclopentadiene
543-27-1	Isobutyl chloroformate [see Chloroformic acid butyl ester]
545-06-2	Trichloroacetonitrile
545-55-1	tris(1-Aziridinyl)-phosphine oxide

546-93-0	Magnesite
551-74-6	Mannomustine dihydrochloride
552-30-7	Trimellitic anhydride
555-84-0	1-[(5-Nitrofurfurylidene)amino]-2-imidazolidinone
556-52-5	Glycidol (2,3-Epoxy-1-propanol)
556-88-7	Nitroguanidine
557-05-1	Zinc stearate
558-13-4	Carbon tetrabromide
562-10-7	Doxylamine succinate
563-04-2	Trimetacresyl phosphate
563-12-2	Ethion
563-41-7	Semicarbazide hydrochloride
563-47-3	3-Chloro-2-methylpropene
563-80-4	Methyl isopropyl ketone (MIPK)
565-59-3	2,3-Dimethylpentane [see Heptane, isomers]
569-61-9	CI Basic Red 9
581-89-5	2-Nitronaphthalene
583-60-8	o-Methylcyclohexanone
584-02-1	3-Pentanol [see Pentanol, all isomers]
584-84-9	Toluene-2,4-diisocyanate (TDI)
589-34-4	3-Methylhexane [see Heptane, isomers]
590-18-1	cis-2-Butene [see Butenes, all isomers]
590-35-2	2,2-Dimethylpentane [see Heptane, isomers]
591-27-5	3-Aminophenol
591-76-4	2-Methylhexane [see Heptane, isomers]
591-78-6	Methyl n-butyl ketone (2-Hexanone)
592-01-8	Calcium cyanide, as CN
592-34-7	Chloroformic acid butyl ester (n-Butyl chloroformate)
592-41-6	1-Hexene
592-45-0	1,4-Hexadiene
592-62-1	Methylazoxymethanol acetate
593-60-2	Vinyl bromide
593-70-4	Chlorofluoromethane (FC-31)

594-15-0	Tribromochloromethane
594-18-3	Dibromodichloromethane
594-27-4	Tetramethyltin
594-42-3	Perchloromethyl mercaptan
594-72-9	1,1-Dichloro-1-nitroethane
598-55-0	Methyl carbamate
598-56-1	N,N-Dimethylethylamine
598-75-4	D-3-Methyl-2-butanol [see Pentanol, all isomers]
598-78-7	2-Chloropropionic acid
599-79-1	Sulfasalazine
600-14-6	2,3-Pentanedione
600-25-9	1-Chloro-1-nitropropane
601-77-4	N-Nitrosodiisopropylamine
602-60-8	9-Nitroanthracene
602-87-9	5-Nitroacenaphthene
603-34-9	Triphenyl amine
603-35-0	Triphenyl phosphine
604-75-1	Oxazepam
606-20-2	2,6-Dinitrotoluene
607-57-8	2-Nitrofluorene
608-73-1	Hexachlorocyclohexane, technical (t-HCH)
608-93-5	Pentachlorobenzene
609-20-1	2,6-Dichloro-p-phenylenediamine
612-00-0	Dowtherm® Q
612-64-6	N-Nitrosoethylphenylamine
612-83-9	3,3'-Dichlorobenzidine dihydrochloride
613-35-4	N,N'-Diacetylbenzidine
614-00-6	N-Nitrosomethylphenylamine
615-05-4	2,4-Diaminoanisole
615-53-2	N-Methyl-N-nitrosourethane
618-85-9	3,5-Dinitrotoluene
620-11-1	3-Pentyl acetate
621-64-7	N-Nitrosodi-n-propylamine (NDPA)
624-41-9	2-Methylbutyl acetate
624-64-6	trans-2-Butene [see Butenes, all isomers]
624-83-9	Methyl isocyanate
624-92-0	Dimethyl disulfide
625-16-1	4-Pentyl acetate (tert-Amyl acetate)
625-45-6	Methoxyacetic acid
626-17-5	Phthalodinitrile, m-isomer
626-38-0	2-Pentyl acetate (sec-Amyl acetate)
627-12-3	n-Propyl carbamate
627-13-4	n-Propyl nitrate
628-63-7	1-Pentyl acetate (n-Amyl acetate)
628-96-6	Ethylene glycol dinitrate (EGDN)
629-11-8	Hexamethylene glycol
630-08-0	Carbon monoxide
630-20-6	1,1,1,2-Tetrachloroethane
631-61-8	Ammonium acetate
631-64-1	Dibromoacetic acid
632-99-5	Magenta
636-21-5	Toluidine hydrochloride, o-isomer
637-07-0	Clofibrate
637-92-3	Ethyl tert-butyl ether (ETBE)
638-21-1	Phenylphosphine
641-48-5	1,2-Dihydroaceanthrylene
643-79-8	o-Phthalaldehyde
646-06-0	1,3-Dioxolane
650-51-1	Sodium trichloroacetate
680-31-9	Hexamethyl phosphoramide
681-84-5	Methyl silicate
684-16-2	Hexafluoroacetone
684-93-5	N-Nitroso-N-methylurea
690-39-1	1,1,1,3,3,3- Hexafluoropropane
692-49-9	cis-1,1,1,4,4,4-Hexafluoro-2-butene
693-21-0	Diethylene glycol dinitrate

693-98-1	2-Methylimidazole
695-77-2	Tetrachlorocyclopentadiene
700-13-0	Trimethylhydroquinone
712-68-5	2-Amino-5-(5-nitro-2-furyl)-1,3,4-thiadiazole
723-46-6	Sulfamethoxazole
730-40-5	Disperse Orange 3
754-12-1	2,3,3,3-Tetrafluoropropene
759-73-9	N-Nitroso-N-ethylurea
763-69-9	Ethyl-3-ethoxypropionate
764-41-0	1,4-Dichloro-2-butene
765-34-4	Glycidaldehyde
768-52-5	N-Isopropylaniline
789-07-1	2-Nitropyrene [see Nitropyrenes]
793-24-8	N-(1,3-Dimethylbutyl)-N′-phenyl-p-phenylene-diamine
794-93-4	Dihydroxymethylfuratrizine
800-24-8	Aziridyl benzoquinone
804-36-4	Nitrovin
811-97-2	1,1,1,2-Tetrafluoroethane (HFC 134a)
817-09-4	Trichlormethine (Trimustine hydrochloride)
818-61-1	2-Hydroxyethyl acrylate
822-06-0	1,6-Hexamethylene diisocyanate
822-36-3	4-Methylimidazole
828-00-2	Dimethoxane
832-69-9	2-Methylphenanthrene
838-88-0	4,4′-Methylene bis(2-methylaniline)
842-07-9	Sudan I
846-50-4	Temazepam
868-77-9	Ethylene glycol methacrylate (2-Hydroxyethyl methacrylate)
868-85-9	Dimethyl hydrogen phosphite
872-05-9	1-Decene
872-50-4	N-Methyl-2-pyrrolidone
892-21-7	3-Nitrofluoranthene
915-67-3	Amaranth
919-86-8	Demeton-S-methyl
920-37-6	2-Chloroacrylonitrile
923-26-2	2-Hydroxypropyl methacrylate (Methacrylic acid 2-hydroxypropyl ester)
924-16-3	N-Nitrosodi-n-butylamine (DBN)
929-06-6	2-(2-Aminoethoxy)ethanol
930-55-2	N-Nitrosopyrrolidine (NPYR)
934-73-6	p-Chlorophenyl methyl sulfoxide
935-92-2	Trimethylquinone
944-22-9	Fonofos
950-37-8	Methidathion
989-38-8	Rhodamine 6G
994-05-8	tert-Amyl methyl ether (TAME)
996-35-0	N,N-Dimethylisopropylamine
998-30-1	Triethoxysilane
999-61-1	2-Hydroxypropyl acrylate
1024-57-3	Heptachlor epoxide
1070-70-8	1,4-Butanediol diacrylate
1072-52-2	2-(1-Aziridinyl)ethanol
1116-54-7	N-Nitrosodiethanolamine (NDELA)
1120-71-4	Propane sultone (1,3-Propane sultone)
1121-03-5	2,4-Butane sultone
1143-38-0	Dithranol
1163-19-5	Decabromodiphenyl oxide
1189-85-1	tert-Butyl chromate, as CrO_3
1239-45-8	Ethidium bromide
1300-73-8	Xylidine, mixed isomers
1302-74-5	Emery
1303-00-0	Gallium arsenide
1303-28-2	Arsenic pentoxide, as As
1303-86-2	Boron oxide
1303-96-4	Sodium tetraborate, decahydrate
1304-82-1	Bismuth telluride, Undoped; Bismuth telluride, Se-doped, as Bi_2Te_3
1305-62-0	Calcium hydroxide

1305-78-8   Calcium oxide
1306-38-3   Cerium oxide and cerium compounds
1307-79-9   Terbufos
1309-37-1   Iron oxide (Fe_2O_3)
1309-48-4   Magnesium oxide
1309-64-4   Antimony trioxide, as Sb
1310-58-3   Potassium hydroxide
1310-65-2   Lithium hydroxide
1310-73-2   Sodium hydroxide
1313-27-5   Molybdenum trioxide
1313-99-1   Nickel oxide
1314-06-3   Nickel sesquioxide
1314-13-2   Zinc oxide; Zinc oxide, fume
1314-56-3   Phosphorus pentoxide
1314-61-0   Tantalum oxide, dusts, as Ta
1314-62-1   Vanadium pentoxide, as V
1314-80-3   Phosphorus pentasulfide
1317-43-7   Nemalite, fibrous dust
1317-60-8   Hematite
1317-65-3   Calcium carbonate (Limestone; Marble)
1317-95-9   Silica, crystalline, tripoli
1318-02-1   Zeolites, excluding erionite
1319-77-3   Cresol, all isomers
1321-64-8   Pentachloronaphthalene
1321-65-9   Trichloronaphthalene
1321-74-0   Divinylbenzene
1327-33-9   Antimony oxide
1327-53-3   Arsenic trioxide, as As
1330-20-7   Xylene (Dimethylbenzene)
1330-43-4   Sodium tetraborate, anhydrous
1332-21-4   Asbestos, all forms
1332-58-7   Kaolin
1333-74-0   Hydrogen

1333-86-4   Carbon black
1335-87-1   Hexachloronaphthalene
1335-88-2   Tetrachloronaphthalene
1336-36-3   Polychlorinated biphenyls (PCBs)
1338-16-5   Iron sorbitol-citric acid complex
1338-23-4   Methyl ethyl ketone peroxide
1338-24-5   Naphthenic acids
1344-28-1   Aluminum oxide (α-Alumina)
1344-95-2   Calcium silicate
1345-04-6   Antimony trisulfide
1345-25-1   Iron oxide (FeO)
1395-21-7   Subtilisins, BPN [see Subtilisins]
1401-55-4   Tannic acid and tannins
1402-68-2   Aflatoxins
1445-75-6   Diisopropyl methylphosphonate (DIMP)
1455-21-6   n-Nonyl mercaptan
1464-53-5   Diepoxybutane
1477-55-0   m-Xylene α,α'-diamine
1484-13-5   Vinylcarbazole
1490-04-6   DL-Menthol, raw
1563-66-2   Carbofuran
1565-94-2   Bisphenol A diglycidyl methacrylate
1569-02-4   1-Ethoxy-2-propanol
1569-69-3   Cyclohexylmercaptan
1582-09-8   Trifluralin
1589-47-5   2-Methoxy-1-propanol (Propylene glycol 2-methyl ether)
1615-80-1   1,2-Diethylhydrazine
1633-83-6   1,4-Butane sultone
1634-04-4   Methyl tert-butyl ether (MTBE)
1639-09-4   n-Heptyl mercaptan (1-Heptanethiol)
1645-83-6   1,3,3,3-Tetrafluoropropylene
1675-54-3   Bisphenol A diglycidylether (4,4'-Isopropylidenediphenol diglycidyl ether)

1680-21-3	Triethylene glycol diacrylate
1689-82-3	4-Hydroxyazobenzene
1694-09-3	Benzyl Violet 4B
1705-85-7	6-Methylchrysene
1706-01-0	3-Methylfluoranthene
1717-00-6	1,1-Dichloro-1-fluoroethane
1738-25-6	Dimethylaminopropionitrile
1746-01-6	2,3,7,8-Tetrachlorodibenzo-p-dioxin (TCDD)
1763-23-1	Perfluorooctanesulfonic acid (PFOS), and its salts
1817-47-6	Nitrocumene, p-isomer
1832-54-8	Isopropyl methyl phosphonic acid (IMPA)
1836-75-5	Nitrofen
1854-23-5	4,4'-(2-Ethyl-2-nitro-1,3-propanediyl)bis morpholine (20% w/v) [see 4-(2-Nitrobutyl) morpholine (70% w/v)/4,4'-(2-Ethyl-2-nitro-1,3-propanediyl)bis morpholine (20% w/v) mixture]
1854-26-8	Dimethyloldihydroxyethylene urea
1897-45-6	Chlorothalonil
1910-42-5	Paraquat dichloride
1912-24-9	Atrazine
1918-02-1	Picloram
1929-82-4	Nitrapyrin (2-Chloro-6-[trichloromethyl]pyridine)
1936-15-8	Orange G
1954-28-5	Triethylene glycol diglycidyl ether
2039-87-4	Chlorostyrene, o-isomer
2050-47-7	p,p'-Dibromodiphenyl ether
2068-78-2	Vincristine sulfate
2074-50-2	Paraquat methosulfate (Paraquat dimethyl sulfate)
2082-79-3	3,5-Di-tert-butyl-4-hydroxyphenyl propionic acid octadecyl ester
2082-81-7	1,4-Butanediol dimethacrylate
2095-03-6	Bisphenyl F diglycidyl ether (p,p'-isomer)
2104-64-5	EPN (O-Ethyl O-[4-nitrophenyl]phenylthiophosphonate)
2164-17-2	Fluometuron
2168-68-5	bis(1-Aziridinyl)morpholinophosphine sulfide (Morzid)
2179-59-1	Allyl propyl disulfide
2210-79-9	Cresyl glycidyl ether, o-isomer
2216-51-5	L-Menthol, natural or synthetic
2224-44-4	4-(2-Nitrobutyl)-morpholine (70% w/v) [see 4-(2-Nitrobutyl) morpholine (70% w/v)/4,4'-(2-Ethyl-2-nitro-1,3-propanediyl)bis morpholine (20% w/v) mixture]
2234-13-1	Octachloronaphthalene
2238-07-5	Diglycidyl ether (DGE)
2243-62-1	1,5-Naphthalenediamine
2303-16-4	Diallate
2318-18-5	Senkirkine
2353-45-9	Fast Green FCF
2358-84-1	Diethylene glycol dimethacrylate
2372-82-9	N-(3-Aminopropyl)-N-dodecylpropane-1,3-diamine
2381-21-7	1-Methylpyrene
2385-85-5	Mirex
2386-87-0	3,4-Epoxycyclohexylmethyl-3,4-epoxycyclohexylcarboxylate
2386-90-5	bis(2,3-Epoxycyclopentyl)ether
2402-79-1	2,3,5,6-Tetrachloropyridine
2425-06-1	Captafol
2425-79-8	1,4-Butanediol diglycidyl ether
2425-85-6	CI Pigment Red 3
2426-08-6	n-Butyl glycidyl ether (BGE)
2429-74-5	CI Direct Blue 15
2431-50-7	2,3,4-Trichloro-1-butene
2432-99-7	11-Aminoundecanoic acid
2451-62-9	1,3,5-Triglycidyl-s-triazinetrione
2455-24-5	Tetrahydrofurfuryl methacrylate
2465-27-2	Auramine hydrochloride
2475-45-8	Disperse Blue 1
2487-90-3	Trimethoxysilane
2527-58-4	2,2'-Dithiobis(N-methylbenzamide)
2528-36-1	Dibutyl phenyl phosphate

2551-62-4	Sulfur hexafluoride
2634-33-5	1,2-Benzisothiazol-3(2H)-one
2646-17-5	Oil Orange SS
2682-20-4	2-Methyl-2,3-dihydroisothiazol-3-one
2687-91-4	N-Ethylpyrrolidone
2698-41-1	Chlorobenzylidene malononitrile, o-isomer
2699-79-8	Sulfuryl fluoride
2757-90-6	Agaritine
2764-72-9	Diquat
2783-94-0	Sunset Yellow FCF
2784-94-3	HC Blue No. 1
2807-30-9	2-Propoxyethanol (Ethylene glycol mono-n-propyl ether)
2832-19-1	Chloroacetamide-N-methylol (CAM)
2832-40-8	Disperse Yellow 3
2835-39-4	Allyl isovalerate
2837-89-0	2-Chloro-1,1,1,2-tetrafluoroethane
2855-13-2	3-Aminomethyl-3,5,5-trimethyl cyclohexylamine (Isophorone diamine)
2867-47-2	N,N'-(Dimethylamino)ethyl methacrylate
2871-01-4	HC Red No. 3
2872-52-8	Disperse Red 1
2885-00-9	Octadecyl mercaptan
2917-26-2	Cetylmercaptan (1-Hexadecanethiol)
2921-88-2	Chlorpyrifos
2955-38-6	Prazepam
2971-90-6	Clopidol
2973-10-6	Diisopropyl sulfate
3018-12-0	Dichloroacetonitrile
3033-62-3	bis(2-Dimethylaminoethyl)ether (DMAEE)
3033-77-0	Glycidyl trimethyl ammonium chloride
3068-88-0	β-Butyrolactone
3101-60-8	p-tert-Butyl phenol glycidyl ether
3118-97-6	Sudan II
3165-93-3	p-Chloro-o-toluidine hydrochloride
3173-72-6	1,5-Naphthylene diisocyanate (NDI)
3179-89-3	Disperse Red 17
3252-43-5	Dibromoacetonitrile
3290-92-4	Trimethylolpropane trimethacrylate
3296-90-0	2,2-bis-(Bromomethyl)-1,3-propanediol, technical grade
3333-52-6	Tetramethyl succinonitrile
3351-28-8	1-Methylchrysene
3351-30-2	4-Methylchrysene
3351-31-3	3-Methylchrysene
3351-32-4	2-Methylchrysene
3383-96-8	Temephos
3522-94-9	2,2,5-Trimethylhexane
3524-68-3	Pentaerythritol triacrylate
3564-09-8	Ponceau 3R
3567-69-9	Carmoisine
3570-75-0	Nifurthiazole (2-[2-Formylhydrazino]-4-[5-nitro-2-furyl]thiazole)
3687-31-8	Lead arsenate
3688-53-7	Furylfuramide
3689-24-5	Sulfotepp (TEDP)
3697-24-3	5-Methylchrysene
3710-84-7	N,N-Diethylhydroxylamine (DEHA)
3761-53-3	Ponceau MX
3771-19-5	Nafenopin
3778-73-2	Isophosphamide
3795-88-8	Levofuraltadone (5-[Morpholinomethyl]-3-[(5-nitrofurfurylidene)amino]-2-oxazolidinone)
3811-73-2	Sodium pyrithione
3825-26-1	Ammonium perfluorooctanoate
3844-45-9	Brilliant Blue FCF, disodium salt
3902-71-4	4,5',8-Trimethylpsoralen
3926-62-3	Sodium chloroacetate
4016-14-2	Isopropyl glycidyl ether (IGE)

4063-41-6	4,5'-Dimethylangelicin plus ultraviolet A radiation
4074-88-8	Diethylene glycol diacrylate
4080-31-3	Methenamin 3-chlor-allylchloride
4098-71-9	Isophorone diisocyanate
4170-30-3	Crotonaldehyde
4342-03-4	Dacarbazine
4548-53-2	Ponceau SX
4549-40-0	N-Nitrosomethylvinylamine
4602-84-0	Farnesol
4657-93-6	5-Aminoacenaphthene
4680-78-8	Guinea Green B
4685-14-7	Paraquat
4687-94-9	Bisphenol A glycerolate
4719-04-4	N,N',N''-tris(β-Hydroxyethyl)-hexahydro-1,3,5-triazine
5026-74-4	Triglycidyl-p-aminophenol
5124-30-1	Methylene bis(4-cyclohexylisocyanate)
5131-60-2	4-Chloro-m-phenylenediamine
5141-20-8	Light Green SF
5160-02-1	D and C Red No. 9 (5-Chloro-2-[(2-hydroxy-1-naphthalenyl)azo]-4-methylbenzenesulfonic acid, barium salt (2:1))
5216-25-1	Chlorobenzotrichloride, p-isomer
5307-14-2	2-Nitro-p-phenylenediamine (1,4-Diamino-2-nitrobenzene)
5332-52-5	1-Undecanethiol (Undecyl mercaptan)
5332-73-0	3-Methoxypropylamine
5385-75-1	Dibenzo[a,e]fluoranthene
5392-40-5	Citral
5431-33-4	Glycidyl oleate
5436-43-1	2,2',4,4'-Tetrabromodiphenyl ether (BDE-47)
5456-28-0	Ethyl selenac
5493-45-8	Hexahydrophthalic acid diglycidyl ester
5522-43-0	1-Nitropyrene [see Nitropyrenes]
5589-96-8	Bromodichloroacetic acid
5714-22-7	Sulfur pentafluoride
5912-86-7	cis-Isoeugenol [see Isoeugenol and its isomers]
5932-68-3	trans-Isoeugenol [see Isoeugenol and its isomers]
5989-27-5	D-Limonene
5989-54-8	L-Limonene (β-Limonene)
6032-29-7	2-Pentanol [see Pentanol, all isomers]
6055-19-2	Cyclophosphamide
6108-10-7	ϵ-Hexachlorocyclohexane
6164-98-3	Chlordimeform
6358-53-8	Citrus Red No. 2
6358-64-1	2,5-Dimethoxy-4-chloroaniline
6368-72-5	Sudan Red 7B
6373-74-6	CI Acid Orange 3
6385-62-2	Diquat dibromide monohydrate [see Diquat]
6416-57-5	Sudan Brown RR
6419-19-8	Aminotris(methylenephosphonic acid)
6423-43-4	Propylene glycol dinitrate (PGDN)
6440-58-0	1,3-Dimethylol-5,5-dimethyl hydantoin
6459-94-5	CI Acid Red 114
6870-67-3	Jacobine
6923-22-4	Monocrotophos
7085-85-0	Ethyl cyanoacrylate (Ethyl 2-cyanoacrylate)
7099-43-6	5,6-Cyclopenteno-1,2-benzanthracene
7411-49-6	3,3'-Diaminobenzidine tetrahydrochloride
7429-90-5	Aluminum, metal and insoluble compounds
7439-92-1	Lead and inorganic compounds, as Pb
7439-93-2	Lithium
7439-96-5	Manganese, and inorganic compounds, as Mn; Manganese, fume, as Mn
7439-97-6	Mercury, aryl compounds, as Hg; Mercury, elemental and inorganic compounds, as Hg
7439-98-7	Molybdenum and insoluble compounds, as Mo; Molybdenum, soluble compounds, as Mo
7440-01-9	Neon
7440-02-0	Nickel compounds; Nickel elemental; Nickel insoluble compounds, as Ni; Nickel soluble compounds, as Ni

7440-06-4   Platinum, metal; Platinum, soluble salts, as Pt
7440-07-5   Plutonium
7440-16-6   Rhodium, elemental
7440-21-3   Silicon
7440-22-4   Silver, metal
7440-25-7   Tantalum, metal
7440-28-0   Thallium and soluble compounds, as Tl
7440-31-5   Tin, metal
7440-33-7   Tungsten
7440-36-0   Antimony and compounds, as Sb
7440-37-1   Argon
7440-38-2   Arsenic and inorganic compounds (except arsine), as As
7440-39-3   Barium and soluble compounds, as Ba
7440-41-7   Beryllium and compounds, as Be
7440-42-8   Boron and compounds
7440-43-9   Cadmium and compounds, as Cd; Cadmium and inorganic
            compounds
7440-47-3   Chromium(III) inorganic compounds, as Cr; Chromium metal
7440-48-4   Cobalt and compounds; Cobalt and inorganic compounds, as
            Co; Cobalt with tungsten carbide
7440-50-8   Copper dusts and mists, as Cu; Copper fume, as Cu; Copper
            and its organic compounds
7440-57-5   Gold and inorganic compounds
7440-58-6   Hafnium and compounds, as Hf
7440-59-7   Helium
7440-61-1   Uranium, natural, soluble and insoluble compounds, as U
7440-62-2   Vanadium and inorganic compounds
7440-65-5   Yttrium and compounds, as Y
7440-66-6   Zinc and compounds
7440-67-7   Zirconium, elemental; Zirconium compounds, as Zr; Zirconium
            insoluble compounds; Zirconium soluble compounds
7440-74-6   Indium and compounds, as In
7446-09-5   Sulfur dioxide

7446-27-7   Lead phosphate
7446-34-6   Selenium sulfide
7460-84-6   Glycidyl stearate
7481-89-2   Zalcitabine
7487-94-7   Mercuric chloride
7496-02-8   6-Nitrochrysene
7519-36-0   N-Nitrosoproline
7550-45-0   Titanium tetrachloride
7553-56-2   Iodine
7572-29-4   Dichloroacetylene
7580-67-8   Lithium hydride
7601-54-9   Trisodium phosphate
7601-89-0   Sodium perchlorate [see Perchlorate and perchlorate salts]
7616-94-6   Perchloryl fluoride
7631-90-5   Sodium bisulfite
7637-07-2   Boron trifluoride
7646-85-7   Zinc chloride, fume
7647-01-0   Hydrogen chloride
7647-10-1   Palladium chloride and other bioavailable Pd(II) compounds
7664-38-2   Phosphoric acid
7664-39-3   Hydrogen fluoride, as F
7664-41-7   Ammonia
7664-93-9   Sulfuric acid
7665-72-7   tert-Butyl glycidyl ether
7681-52-9   Sodium hypochlorite
7681-57-4   Sodium metabisulfite
7697-37-2   Nitric acid
7699-41-4   Fused Silica [see Silica, amorphous, fused]
7718-54-9   Nickel chloride
7719-09-7   Thionyl chloride
7719-12-2   Phosphorus trichloride
7722-84-1   Hydrogen peroxide
7722-88-5   Tetrasodium pyrophosphate

CAS Number	Name
7723-14-0	Phosphorus, White
7726-95-6	Bromine
7727-21-1	Potassium persulfate [see Persulfates, as persulfate]
7727-27-1	Sodium persulfate [see Persulfates, as persulfate]
7727-37-9	Nitrogen
7727-43-7	Barium sulfate
7727-54-0	Ammonium persulfate, as S_2O_8
7738-94-5	Chromic acid and chromates
7747-35-5	5-Ethyl-3,7-dioxa-1-azabicyclo[3.3.0]octane
7758-01-2	Potassium bromate
7758-19-2	Chlorite, sodium salt
7758-97-6	Lead chromate
7773-06-0	Ammonium sulfamate
7775-27-1	Sodium persulfate, as S_2O_8
7778-18-9	Calcium sulfate
7778-39-4	Arsenic acid, and its salts, as As
7778-44-1	Calcium arsenate, as As
7778-74-7	Potassium perchlorate [see Perchlorate and perchlorate salts]
7779-27-3	1,3,5-Triethylhexahydro-1,3,5-triazine
7782-41-4	Fluorine
7782-42-5	Graphite, natural
7782-49-2	Selenium compounds, as Se; Selenium, inorganic compounds, as Se; Selenium metal
7782-50-5	Chlorine
7782-65-2	Germanium tetrahydride
7782-79-8	Hydrazoic acid
7783-00-8	Selenious acid
7783-06-4	Hydrogen sulfide
7783-07-5	Hydrogen selenide
7783-41-7	Oxygen difluoride
7783-54-2	Nitrogen trifluoride
7783-60-0	Sulfur tetrafluoride
7783-79-1	Selenium hexafluoride, as Se
7783-80-4	Tellurium hexafluoride, as Te
7784-42-1	Arsine
7786-34-7	Mevinphos
7786-81-4	Nickel sulfate
7789-06-2	Strontium chromate, as Cr
7789-30-2	Bromine pentafluoride
7790-91-2	Chlorine trifluoride
7790-94-5	Chlorosulfonic acid
7790-98-9	Perchlorate and perchlorate salts
7791-03-9	Lithium perchlorate, anhydrous [see Perchlorate and perchlorate salts]
7803-49-8	Hydroxylamine (and its salts)
7803-51-2	Phosphine
7803-52-3	Antimony hydride (Stibine)
7803-57-8	Hydrazine hydrate and hydrazine salts
7803-62-5	Silicon tetrahydride (Silane)
8001-35-2	Chlorinated camphene (Toxaphene)
8001-50-1	Terpene polychlorinates
8001-58-9	Creosotes
8002-05-9	Petroleum distillates, Naphtha (Rubber solvent)
8002-26-4	Tall oil, distilled
8002-74-2	Paraffin wax fume
8003-34-7	Pyrethrum
8004-13-5	Phenyl ether/biphenyl mixture, vapor
8006-61-9	Gasoline
8006-64-2	Turpentine [see Turpentine and selected monoterpenes]
8008-20-6	Kerosene [see Kerosene/Jet fuels as total hydrocarbon vapor]
8012-95-1	Oil mist, mineral
8018-01-7	Mancozeb
8018-07-3	Acriflavinium chloride
8022-00-2	Methyl demeton (Demeton-methyl)
8030-30-6	Naphtha, coal tar
8032-32-4	VM & P naphtha

8042-47-5	White mineral oil
8047-67-4	Saccharated iron oxide
8050-09-7	Rosin core solder thermal decomposition products (Colophony)
8052-41-3	Stoddard solvent
8052-42-4	Asphalt fume (Bitumen)
8065-48-3	Demeton
9000-07-1	Carrageenan, native
9001-00-7	Bromelain
9001-73-4	Papain
9001-75-6	Pepsin
9002-07-7	Trypsin
9002-84-0	Polytetrafluoroethylene
9002-86-2	Polyvinyl chloride (PVC)
9002-88-4	Polyethylene
9002-89-5	Polyvinyl alcohol
9003-01-4	Polyacrylic acid
9003-04-7	Acrylic acid polymer, neutralized, cross-linked
9003-07-0	Polypropylene
9003-20-7	Polyvinyl acetate
9003-22-9	Vinyl chloride–Vinyl acetate copolymers
9003-31-0	Natural rubber latex, as inhalable allergenic proteins
9003-39-8	Polyvinyl pyrrolidone
9003-53-6	Polystyrene
9003-54-7	Styrene-acrylonitrile copolymers
9003-55-8	Styrene-butadiene copolymers
9004-07-3	Chymotrypsin
9004-34-6	Cellulose
9004-51-7	Iron-dextrin complex
9004-66-4	Iron-dextran complex
9005-25-8	Starch
9006-04-6	Natural rubber latex, as inhalable allergenic proteins
9009-54-5	Polyurethane foams
9010-98-4	Polychloroprene

9011-06-7	Vinylidene chloride–Vinyl chloride copolymers
9011-14-7	Polymethyl methacrylate
9014-01-1	Subtilisin Carlsberg [see Subtilisins]
9016-87-9	Polymethylene polyphenyl isocyanate (Polymeric MDI)
10024-97-2	Nitrous oxide
10025-67-9	Sulfur monochloride
10025-78-2	Trichlorosilane
10025-87-3	Phosphorus oxychloride
10026-04-7	Tetrachlorosilane
10026-13-8	Phosphorus pentachloride
10028-15-6	Ozone
10034-76-1	Calcium sulfate hemihydrate [see Calcium sulfate]
10034-93-2	Hydrazine sulfate
10035-10-6	Hydrogen bromide
10043-35-3	Boric acid [see Borate compounds, inorganic]
10048-13-2	Sterigmatocystin
10048-32-5	Parasorbic acid
10049-04-4	Chlorine dioxide
10101-41-4	Calcium sulfate dihydrate [see Calcium sulfate]
10102-43-9	Nitric oxide
10102-44-0	Nitrogen dioxide
10124-43-3	Cobalt sulfate
10210-68-1	Cobalt carbonyl, as Co
10222-01-2	2,2-Dibromo-2-cyanacetamide
10294-33-4	Boron tribromide
10294-34-5	Boron trichloride
10380-28-6	Copper-8-hydroxyquinoline
10540-29-1	Tamoxifen
10595-95-6	N-Nitrosomethylethylamine
10599-90-3	Chloramine
10605-21-7	Carbendazim
11056-06-7	Bleomycins
11070-44-3	Methyltetrahydrophthalic anhydride

11097-69-1	Chlorodiphenyl, 54% chlorine
11103-86-9	Potassium zinc chromate hydroxide [see Zinc chromates, as Cr]
11107-01-0	Tungsten carbide, mixed with Co (85% WC : 15% Co)
12001-26-2	Mica
12001-28-4	Crocidolite [see Asbestos, all forms]
12001-79-5	Vitamin K substances
12011-76-6	Dawsonite, fibrous dust
12035-36-8	Nickel dioxide
12035-72-2	Nickel subsulfide
12054-48-7	Nickel hydroxide
12057-24-8	Lithium oxide
12070-12-1	Tungsten carbide
12079-65-1	Manganese cyclopentadienyl tricarbonyl, as Mn
12108-13-3	2-Methylcyclopentadienyl manganese tricarbonyl, as Mn
12122-67-7	Zineb
12125-02-9	Ammonium chloride fume
12172-73-5	Amosite [see Asbestos, all forms]
12174-11-7	Attapulgite, fibrous dust (Palygorskite)
12179-04-3	Sodium tetraborate, pentahydrate
12185-10-3	Phosphorus, Yellow
12192-57-3	Aurothioglucose
12286-12-3	Magnesium oxide sulfate, fibrous dust
12298-43-0	Halloysite, fibrous dust
12427-38-2	Manganous ethylenebis(dithiocarbamate) (Maneb)
12510-42-8	Erionite, fibrous dust
12604-58-9	Ferrovanadium dust
12663-46-6	Cyclochlorotine
12718-69-3	Tungsten carbide, mixed with Co (92% WC : 8% Co)
12789-03-6	Chlordane, technical grade [see Chlordane]
13045-94-8	Medphalan
13048-33-4	1,6-Hexanediol diacrylate
13010-47-4	1-(2-Chloroethyl)-3-cyclohexyl-1-nitrosourea (CCNU)
13121-70-5	Cyhexatin (Tricyclohexyltin hydroxide)
13149-00-3	Hexahydrophthalic anhydride, cis isomer [see Hexahydrophthalic anhydride, all isomers]
13256-22-9	N-Nitrososarcosine
13292-46-1	Rifampicin
13360-57-1	Dimethylsulfamoyl chloride
13397-24-5	Gypsum [see Calcium sulfate]
13463-39-3	Nickel carbonyl, as Ni
13463-40-6	Iron pentacarbonyl
13463-41-7	Zinc pyrithione
13463-67-7	Titanium dioxide
13464-58-9	Arsenous acid and its salts, as As
13466-78-9	Δ-3-Carene [see Turpentine and selected monoterpenes]
13483-18-6	1,2-Bis(chloromethoxy)ethane
13494-80-9	Tellurium and compounds, as Te
13530-65-9	Zinc chromates, as Cr
13552-44-8	4,4′-Methylenedianiline dihydrochloride
13756-19-0	Calcium chromate, as Cr
13838-16-9	Enflurane
13909-09-6	1-(2-Chloroethyl)-3-(4-methylcyclohexyl)-1-nitrosourea (Methyl-CCNU; Semustine)
13952-84-6	sec-Butylamine
13983-17-0	Wollastonite
14166-21-3	Hexahydrophthalic anhydride, trans-isomer [see Hexahydrophthalic anhydride, all isomers]
14464-46-1	Silica, crystalline, cristobalite
14484-64-1	Ferbam
14548-60-8	Benzylhemiformal
14596-37-3	Phosphorus-32
14807-96-6	Talc, containing no asbestos fibers
14808-60-7	Silica, crystalline, α-quartz
14857-34-2	Dimethylethoxysilane
14861-17-7	4-(2,4-Dichlorophenoxy)benzenamine
14901-08-7	Cycasin

14977-61-8 Chromyl chloride
15086-94-9 Eosin
15096-52-3 Sodium aluminum fluoride, as F
15141-18-1 Disperse Blue 124 [see Disperse Blue 106/124]
15159-40-7 N-Chloroformylmorpholine
15356-70-4 DL-Menthol, synthetic
15468-32-3 Silica, crystalline, tridymite
15501-74-3 Sepiolite, fibrous dust
15503-86-3 Isatidine
15541-45-4 Bromate
15625-89-5 Trimethylolpropane triacrylate
15627-09-5 N-Cyclohexylhydroxydiazene-1-oxide, copper salt
15663-27-1 Cisplatin
15721-02-5 2,2′,5,5′-Tetrachlorobenzidine
15922-78-8 Sodium pyrithione
15972-60-8 Alachlor
16065-83-1 Chromium(III)
16096-31-4 Diglycidyl hexanediol
16219-75-3 Ethylidene norbornene
16543-55-8 4-(N-Nitrosomethylamino)-4-(3-pyridyl)-1-butanal (NNA)
16568-02-8 Gyromitrin
16752-77-5 Methomyl
16812-54-7 Nickel sulfide
16842-03-8 Cobalt hydrocarbonyl, as Co
17117-34-9 3-Nitrobenzanthrone
17702-41-9 Decaborane
17804-35-2 Benomyl
17831-71-9 Tetraethylene glycol diacrylate
18282-10-5 Stannic oxide [see Tin oxides, as Sn]
18307-23-8 Sepiolite
18454-12-1 Lead chromate oxide
18540-29-9 Chromium(VI)
18883-66-4 Streptozotocin

19044-88-3 Oryzalin
19287-45-7 Diborane
19408-74-3 1,2,3,7,8,9-Hexachlorodibenzo-p-dioxin [see Hexachlorodibenzo-p-dioxin, mixture (HxCDD)]
19430-93-4 Perfluorobutyl ethylene (PFBE)
19624-22-7 Pentaborane
20073-24-9 3-Carbethoxypsoralen
20268-51-3 7-Nitrobenz[a]anthracene
20589-63-3 3-Nitroperylene
20706-25-6 2-Propoxyethyl acetate (Ethylene glycol monopropyl ether acetate)
20816-12-0 Osmium tetroxide
20830-75-5 Digoxin
20830-81-3 Daunomycin
20941-65-5 Ethyl telluric
21087-64-9 Metribuzin
21351-79-1 Cesium hydroxide
21645-51-2 Aluminum hydroxide
21651-19-4 Tin oxide, as Sn
21725-46-2 Cyanazine
21884-44-6 Luteoskyrin
22224-92-6 Fenamiphos
22248-79-9 Tetrachlorvinphos
22349-59-3 1,4-Dimethylphenanthrene
22398-80-7 Indium phosphide
22506-53-2 3,9-Dinitrofluoranthene
22571-95-5 Symphytine
22781-23-3 Bendiocarb
22966-79-6 Estradiol mustard
22967-92-6 Methyl mercury
22975-76-4 4,4′-Dimethylangelicin plus ultraviolet A radiation
23134-05-6 Metabisulfites
23209-59-8 Calcium sodium metaphosphate (fibrous dust)
23214-92-8 Adriamycin®, Doxorubicin hydrochloride

CAS Number	Name
23246-96-0	Riddelliine
23255-93-8	Hycanthone mesylate
23537-16-8	Rugulosin
23696-28-8	Olaquindox
23746-34-1	Potassium bis(2-hydroxyethyl)dithiocarbamate
24124-25-2	Tributyltin linoleate [see n-Butyltin compounds, as Sn]
24448-20-2	Bisphenol A ethoxylate dimethacrylate
24560-98-3	9,10-Epoxystearic acid, cis-isomer
24938-64-5	Kevlar 49 (p-Aramid fibrils)
25013-15-4	Vinyl toluene
25013-16-5	Butylated hydroxyanisole (BHA)
25038-54-4	Nylon 6
25057-89-0	Bentazon (Basagran)
25154-54-5	Dinitrobenzene, all isomers
25167-67-3	Butene, mixture of isomers [see Butenes, all isomers]
25265-71-8	Dipropylene glycol
25321-14-6	Dinitrotoluene
25322-68-3	Polyethylene glycol(s)
25322-69-4	Polypropylene glycol(s)
25329-35-5	Pentachlorocyclopentadiene
25340-17-4	Diethylbenzene, mixed isomers (Dowtherm® J)
25376-45-8	2,4-Toluenediamine, mixed isomers [see 2,4-Diaminotoluene]
25551-13-7	Trimethylbenzene, all isomers
25584-83-2	Hydroxypropyl acrylate, all isomers (Acrylic acid hydroxypropyl ester)
25639-42-3	Methylcyclohexanol
25732-74-5	Acepyrene
25812-30-0	Gemfibrozil
25962-77-0	trans-2-[(Dimethylamino)methylimino]-5-[2-(5-nitro-2-furyl)-vinyl]-1,3,4-oxadiazole
26125-61-1	p-Aramide, fibrous dust
26140-60-3	Terphenyl, mixture [see Terphenyl, o-, m-, p-isomers]
26148-68-5	2-Amino-9H-pyrido[2,3-b]indole (A-α-C)
26172-55-4	5-Chloro-2-methyl-2,3-dihydroisothiazol-3-one
26308-28-1	Ripazepam
26447-14-3	Cresyl glycidyl ether, mixture of isomers
26471-62-5	Toluene diisocyanate, mixed isomers [see Toluene-2,4-diisocyanate]
26499-65-0	Plaster of Paris (Calcium sulfate hemihydrate)
26530-20-1	2-Octyl-4-isothiazolin-3-one
26628-22-8	Sodium azide
26636-01-1	Dimethyltin bis(isooctylmercaptoacetate)
26761-40-0	Diisodecyl phthalate
26782-43-4	Hydroxysenkirkine
26952-21-6	Isooctyl alcohol
27208-37-3	Cyclopenta[cd]pyrene
27478-34-8	Dinitronaphthalene, all isomers
28434-86-8	3,3′-Dichloro-4,4′-diaminodiphenyl ether
28767-61-5	1,3,6,8-Tetranitropyrene [see Nitropyrenes]
28768-32-3	Tetraglycidyl-4,4′-methylenedianiline
29069-24-7	Prednimustine
29118-24-9	1,3,3,3-Tetrafluoropropylene
29171-20-8	2-Dehydrolinalool
29222-48-8	Trimethylpentane, all isomers
29291-35-8	N-Nitrosofolic acid
29590-42-9	Isooctyl acrylate (2-Propenoic acid, isooctyl ester)
29767-20-2	Teniposide
29975-16-4	Estazolam
30310-80-6	N-Nitrosohydroxyproline
30516-87-1	Zidovudine (AZT)
30560-19-1	Acephate
30618-84-9	Glyceryl monothioglycolate
30899-19-5	Pentanol, all isomers
31027-31-3	4-Isopropylphenyl isocyanate
31242-93-0	o-Chlorinated diphenyl oxide [see Chlorinated diphenyl oxide]
31565-23-8	Di(tert-dodecyl)pentasulfide

32534-81-9   Pentabromodiphenyl ether
32536-52-0   Octabromodiphenyl ether
33229-34-4   HC Blue No. 2
33419-42-0   Etoposide
33543-31-6   2-Methylfluoranthene
34590-94-8   (2-Methoxymethylethoxy)propanol (DPGME)
35074-77-2   Hexamethylene bis(3-[3,5-di-*tert*-butyl-4-hydroxyphenyl]
         propionate
35400-43-2   Sulprofos
35554-44-0   1-(2-Allyloxy)-2-(2,4-dichlorophenyl)ethyl-1*H*-imidazole
35691-65-7   1,2-Dibromo-2,4-dicyanobutane
37278-89-0   Xylanases
37300-23-5   Zinc Yellow [*see* Zinc chromates, as Cr]
37329-49-0   Tungsten carbide, mixed with Co and Ti (78% WC : 14% Co :
         8% Ti)
37620-20-5   N′-Nitrosoanabasine (NAB)
37971-36-1   2-Phosphono-1,2,4-butanetricarboxylic acid
38571-73-2   1,2,3-tris(Chloromethoxy)propane
39148-24-8   Fosetyl-al
39156-41-7   2,4-Diaminoanisole sulfate
39413-47-3   Zinc beryllium silicate, as Be
39638-32-9   bis-(2-Chloroisopropyl)ether
40088-47-9   Tetrabromodiphenyl ether
40762-15-0   Doxefazepam
41683-62-9   1,2-Dichloromethoxyethane
41851-50-7   Chlorocyclopentadiene
42397-64-8   1,6-Dinitropyrene
42397-65-9   1,8-Dinitropyrene
42978-66-5   Tripropylene glycol diacrylate
49690-94-0   Tribromodiphenyl ether
51218-45-2   Metolachlor
51264-14-3   Amsacrine
51481-61-9   Cimetidine

51630-58-1   Fenvalerate
52645-53-1   Permethrin
52918-63-5   Deltamethrin
53306-54-0   Di(2-propylheptyl)phthalate
53469-21-9   Chlorodiphenyl, 42% chlorine
53973-98-1   Carrageenan, degraded
54208-63-8   Bisphenyl F diglycidyl ether (o,o′-isomer)
54749-90-5   Chlorozotocin
54839-24-6   1-Ethoxy-2-propyl acetate
54849-38-6   Methyltin tris(isooctylmercaptoacetate)
55290-64-7   Dimethipin
55406-53-6   3-Iodo-2-propynyl butylcarbamate
55557-01-2   N-Nitrosoguvacine
55557-02-3   N-Nitrosoguvacoline
55566-30-8   Tetrakis(hydroxymethyl)phosphonium sulfate
55720-99-5   Chlorinated diphenyl oxide
56894-91-8   1,4-bis(Chloromethoxymethyl)benzene
57018-52-7   1-tert-Butoxy-2-propanol
57117-31-4   2,3,4,7,8-Pentachlorodibenzofuran
57465-28-8   3,4,5,3′,4′-Pentachlorobiphenyl (PCB-126)
57469-07-5   Bisphenyl F diglycidyl ether (o,p′-isomer)
57583-35-4   Dimethyltin bis(2-ethylhexylmercaptoacetate)
57653-85-7   1,2,3,6,7,8-Hexachlorodibenzo-p-dioxin [*see* Hexachlorodibenzo-
         p-dioxin, mixture]
57835-92-4   4-Nitropyrene [*see* Nitropyrenes]
59118-99-9   bis[Methyltin di(2-mercaptoethyloleate)]sulfide
59277-89-3   Aciclovir
59536-65-1   Polybrominated biphenyls (PBBs)
59820-43-8   HC Yellow No. 4
59865-13-3   Cyclosporin A
60102-37-6   Pentasitenine
60153-49-3   3-(Methylnitrosamino)propionitrile
         (3-[*N*-Nitrosomethylamino]propionitrile)

60348-60-9	2,2′,4,4′,5-Pentabromodiphenyl ether (BDE-99)
60568-05-0	Furmecyclox
60676-86-0	Silica, amorphous, fused
61788-32-7	Hydrogenated terphenyls
61789-36-4	Calcium naphthenate [see Naphthenate, Na-, Ca-, K-]
61789-86-4	Petroleum sulfonates, calcium salts
61790-13-4	Sodium naphthenate [see Naphthenate, Na-, Ca-, K-]
61790-53-2	Silica, amorphous, diatomaceous earth, uncalcined
61951-51-7	Disperse Blue 124 [see Disperse Blue 106/124]
62450-06-0	3-Amino-1,4-dimethyl-5H-pyrido[4,3-b]indole (Trp-P-1)
62450-07-1	3-Amino-1-methyl-5H-pyrido[4,3-b]indole (Trp-P-2)
62765-93-9	NIAX® Catalyst ESN
63021-86-3	Nitropyrene [see Nitropyrenes]
63041-90-7	6-Nitrobenzo[a]pyrene
63936-56-1	Nonabromodiphenyl ether
64091-91-4	4-(N-Nitrosomethylamino)-1-(3-pyridyl)-1-butanone
64742-47-8	Petroleum distillates, hydrotreated light
64742-48-9	Naphtha, petroleum, hydrotreated, heavy
64742-81-0	Hydrosulfurized kerosene [see Kerosene/Jet fuels as total hydrocarbon vapor]
65271-80-9	Mitoxantrone
65996-93-2	Coal tar pitch volatiles, as benzene soluble aerosol
65997-15-1	Portland cement
66072-08-0	Potassium naphthenate [see Naphthenate, Na-, Ca-, K-]
66204-44-2	N,N′-Methylene-bis(5-methyloxazolidine)
66603-10-9	Cyclohexylhydroxydiazene-1-oxide, potassium salt
66733-21-9	Erionite, fibrous dust
67730-10-3	2-Aminodipyrido[1,2-a:3′,2′-d]imidazole (Glu-P-2)
67730-11-4	2-Amino-6-methyldipyrido[1,2-a:3′,2′-d]imidazole (Glu-P-1)
67747-09-5	Prochloraz
68006-83-7	2-Amino-3-methyl-9H-pyrido[2,3-b]indole(MeA-α-C)
68308-34-9	Shale-oils
68334-30-5	Diesel oil [see Diesel fuel]

68359-37-5	Cyfluthrin
68425-15-0	Polysulfides, di-tert-dodecyl [see Di(tert-dodecyl)polysulfides]
68476-30-2	Fuel oil No. 2 [see Diesel fuel]
68476-34-6	Diesel fuel No. 2 [see Diesel fuel]
68476-85-7	L.P.G. (Liquefied petroleum gas)
68516-81-4	Disperse Blue 106/124
68583-56-2	tert-Dodecyl mercaptan, sulfur reaction product [see Di(tert-dodecyl)polysulfides]
68603-42-9	Cocamide diethanolamine
68895-54-9	Silica, amorphous, diatomaceous earth, calcined
68916-96-1	Mate, absolute (Tea oil)
68937-41-7	Triphenyl phosphate isopropylated
68987-42-8	Dowtherm® Q
69012-64-2	Silica, amorphous, silica fume
69655-05-6	Didanosine
70657-70-4	2-Methoxypropyl-1-acetate (Propylene glycol 2-methyl ether-1-acetate)
71267-22-6	N′-Nitrosoanatabine (NAT)
72178-02-0	Fomesafen
73459-03-7	5-Methylangelicin plus ultraviolet A radiation
74115-24-5	Apollo
74222-97-2	Sulfometuron methyl
75321-19-6	1,3,6-Trinitropyrene [see Nitropyrenes]
75321-20-9	1,3-Dinitropyrene
76180-96-6	2-Amino-3-methylimidazo[4,5-f]quinoline (IQ)
76578-14-8	Assure
77094-11-2	2-Amino-3,4-dimethylimidazo[4,5-f]quinoline (MeIQ)
77323-84-3	Trichlorocyclopentadiene
77439-76-0	3-Chloro-4-(dichloromethyl)-5-hydroxy-2(5H)-furanone
77500-04-0	2-Amino-3,8-dimethylimidazo[4,5-f]quinoxaline (MeIQx)
77536-66-4	Actinolite [see Asbestos, all forms]
77536-67-5	Anthophyllite [see Asbestos, all forms]
77536-68-6	Tremolite [see Asbestos, all forms]

77650-28-3   Diesel fuel, marine [see Diesel fuel]
78432-19-6   Dinitropyrene [see Nitropyrenes]
79217-60-0   Cyclosporine
80508-23-2   N′-Nitrosonornicotine (NNN)
82413-20-5   Droloxifene
82558-50-7   Isoxaben
83463-62-1   Bromochloroacetonitrile
85502-23-4   3-(N-Nitrosomethylamino)propionaldehyde
85878-62-2   Pyrido[3,4-c]psoralen
85878-63-3   7-Methylpyrido[3,4-c]-psoralen
86290-81-5   Gasoline
87625-62-5   Ptaquiloside
89778-26-7   Toremifene
90370-29-9   4,4′,6-Trimethylangelicin plus ultraviolet A radiation
90456-67-0   N-Methylolacrylamide
91273-04-0   N,N-bis(2-Ethylhexyl)-(1,2,4-triazole-1-yl)methanamine
93763-70-3   Perlite
94624-12-1   Pentanol-1/3-Methyl butanol-1 mixture [see Pentanol, all
              isomers]

95465-99-9   Cadusafos
95481-62-2   Dicarboxylic acid (C$_4$–C$_6$) dimethyl ester, mixture
101043-37-2   Microcystin-LR
105650-23-5   2-Amino-1-methyl-6-phenylimidazo[4,5-b]pyridine [PhIP]
105735-71-5   3,7-Dinitrofluoranthene
111025-46-8   Pioglitazone
111189-32-3   Naphtho[1,2-b]fluoranthene
112926-00-8   Silica, amorphous, precipitated and gel
116355-83-0   Fumonisin B$_1$
118399-22-7   Nodularins
128639-02-1   Carfentrazone-ethyl
131341-86-1   Fludioxonil
132207-32-0   Chrysotile [see Asbestos, all forms]
136677-10-6   Polychlorinated dibenzofurans
163702-07-6   1,1,1,2,2,3,3,4,4-Nonafluoro-4-methoxybutane [see HFE-7100]
163702-08-7   2-(Difluoromethoxymethyl)-1,1,1,2,3,3,3-heptafluoropropane [see
              HFE-7100]